Leckie × Leckie

HIGHER Biology

grade booster

John Di Mambro

01/261208

ISBN 978-1-84372-720-0

Published by
Leckie & Leckie Ltd, 3rd floor, 4 Queen Street, Edinburgh, EH2 1JE
Tel: 0131 220 6831 Fax: 0131 225 9987
enquiries@leckieandleckie.co.uk www.leckieandleckie.co.uk

Special thanks to
Aptara Inc. (design and page makeup), Roda Morrison (copy-editing), Caleb Rutherford (Cover design), Jennifer Shaw (proofreading).

A CIP Catalogue record for this book is available from the British Library.

® Leckie & Leckie is a registered trademark
Leckie & Leckie Ltd is a division of Huveaux plc.

Acknowledgements
We would like to thank the SQA for permission to reproduce past examination questions.

CONTENTS

This book is dedicated to Danny

Introduction

WHAT THIS BOOK IS

This book is designed to help you obtain a better grade in your Higher Biology. It will take you through techniques for answering questions, as well as suggesting study skills which will help you learn theory and practise data-handling. All the questions in the book are taken from actual National Examinations. Remember that Biology is very much a knowledge-based subject and therefore you must have an appropriate and extensive database of information at your fingertips. Learning facts is, of course, common to all your Higher Grade subjects, but it especially underpins Biology. In addition, the application of this knowledge plays an increasingly large part in Biology and is reflected in the variety of problem-solving type questions which will appear in the National Examination.

WHAT THIS BOOK IS NOT

This book is not a set of facts for you to learn. If you need this type of course content-based text, you should seek advice from your teachers and peers as to which are the most suitable.

HOW TO USE THIS BOOK

Interacting with learning material is a powerful way of obtaining feedback on your strengths and weaknesses. The material in this book will give you the opportunity to work on the different strategies needed to handle the Higher examination questions and will help you make the best use of what you know and can do. Hopefully, you will develop an appreciation of areas where you need to work to hone your skills but also pick up on others where you know you are doing well. Asking questions, especially of your teacher, is fundamental to improvement and confidence building so this is something you should be doing on a regular basis.

The book is not meant to be read continuously from cover to cover, though it will serve you very well to have done this by the end of your course. With experience, you will develop a self-awareness of your areas of weakness which require support. You should visit these areas several times using the exemplars here to help.

WHEN TO USE THIS BOOK

As with learning most things, the earlier the better. Once you have started your course, you will rapidly meet questions, typical of the National Examination, relevant to the unit material you are working on. Using this book simultaneously will be of great benefit. The structure of the course and the design of the examination are described below. Read up on these early on and come back to refresh your memory. As the formal school and National Examinations come close, you will want to use this book more intensively to help you understand the typical structure of questions and how best to tackle them. Your aim is to have general and specific skills which can be transferred to different contexts, making best use of what you have learned.

COURSE DESIGN AND ASSESSMENT

You can find out very precise details about the design of the Higher Biology by looking at the Scottish Qualifications Authority (SQA) website at www.sqa.org.uk. You are allowed to print out anything here provided it is not used for any purpose other than for your own use as a student. The outline of the course design detailed below is based on the information on this website.

There are three units, each of which lasts 40 hours, covering Cell Biology, Genetics and Adaptation, and Control and Regulation. Additionally, the course includes 40 hours which can be allocated for basic introduction, support, revision, extension work, examination preparation and so on.

When each of these three units has been completed, it is assessed separately and internally by a test lasting 45 minutes, worth 40 marks. In order to pass the internal assessments you need to score 26 or more marks. As with the National Examinations, there are no half marks in these assessments. However, you should be aware that these tests have only structured questions in them, with no other style of questions such as multiple-choice or extended responses, and are aimed approximately at grade C level. If you do not pass an internal test, you will be given the chance to do further revision before resitting a different test. You will also have to write up one experimental report based on a piece of practical work set by your teacher but subject to strict guidelines given by the SQA. These are all done in school time and assessed by your teacher.

In the National Examinations, usually sometime in May, you will take a formal examination for Higher Biology lasting 2 hours 30 minutes under, by then, familiar examination conditions. This is set and marked externally by the SQA. This examination will test you on the whole course with questions which relate to different parts at the same time. In other words, the questions need not be related to just one part of one unit.

The National Examination will have three sections totalling 130 marks in all.

Section A:	30 multiple-choice questions	30 marks
Section B:	short answer questions	80 marks
Section C:	2 extended responses chosen from 4	20 marks

The questions are designed to test you in two main areas of knowledge and understanding (KU) and problem-solving (PS).

Section A has approximately 9–11 PS questions with the remainder being KU. You have to answer all these questions.

Section B contains structured questions and data-handling questions. Between 25 and 30 marks are PS and/or practical abilities type questions, the remainder being KU. You have to answer all these questions.

Section C has 4 extended responses to test your ability to select, organise and present relevant knowledge. The 20 marks are allocated as follows:

- 2 structured questions each worth 10 marks from which you will choose 1 question.

- 2 unstructured questions each worth 10 marks from which you will chose 1 question. 8 marks are allocated for KU and 1 mark for relevance and 1 for coherence.

Your final award, assuming you have passed all the internal assessments, will be based on how well you perform in this National Examination. You will be awarded a pass at A, B, C or D depending on how well you have done. As a rough guide, you will need to get round 75%, 65%, 55% and 45% to obtain these grades respectively but, remember, this is only a guide. The Principal Examiner for the SQA in 2005 said to achieve an A grade candidates must be able '... to provide good and concise explanations of important biological concepts thus demonstrating a clear understanding. The A grade candidates can also apply their knowledge and understanding effectively in situations or contexts less familiar to them.'

KNOWLEDGE AND UNDERSTANDING

As mentioned earlier, Biology is very knowledge-based and this is reflected in the bias of the marks allocated in all three sections of the National Examination. For KU, you will need to be able to recall and recognise important facts and principles. The course is extensive and needs you to avoid compartmentalising your knowledge. In other words, information on one topic will often need to be used to explain or understand what is happening in another context. For example, knowing how ATP is produced and used, learned in Unit 1, might well be relevant in explaining how energy is provided for active transport or osmoregulation or maintaining body temperature. Only by practice and reflection will you develop the ability to see connections between different parts of the course.

PROBLEM-SOLVING

This is a skill which often causes difficulties for students. Here you are required to be able to:

- select relevant information from texts, tables, charts, keys, graphs and diagrams
- present information appropriately in a variety of forms, including written summaries, extended writing, tables and graphs
- process information accurately, using calculations where appropriate
- plan, design and evaluate experimental procedures

- draw valid conclusions and give explanations supported by evidence
- make predictions and generalisations based on available evidence

It is generally agreed that a powerful way of improving you problem-solving abilities is to practise doing questions based on PS. Of particular relevance here is constant access to the past papers and working through these methodically, flagging up any areas which need strengthening. It is important not to take shortcuts with PS questions which can often lead to loss of marks through careless errors.

TERMINOLOGY

Biology is riddled with its own terminology, sometimes difficult to understand and remember. It is very good practice to list these terms by topic and make sure you fully understand their meaning. Using word processing software, you can easily make up a database of such terms and rearrange them alphabetically, as well as by topic. Such key terms are a powerful way of revising because you can use them in two important ways. Suppose you write a key term on one side of a small index card and the definition on the other. Now, by randomly picking up such a 'flash card' from a stack which you have gathered, you could look at the definition and recall the key term or, alternatively, by looking at the key term, you could try to recall the definition. By making the cards attractive and colourful, you will increase the chances of being able to form strong links between a term and its definition. The time spent on this exercise will pay itself back many fold. You could also make this type of exercise one you share with a friend, testing each other. Also, try not to learn terms just by rote which is one of the most ineffective ways to learn. If you can see how words are constructed, you can sometimes work out what an unfamiliar term means. Suppose you meet the word 'molluscicide' for the first time and think to yourself, 'I've no idea what this means.' Don't panic, think of these words you have met before which sound similar, e.g. pesticide, fungicide, herbicide. Now do you think you can work out what the new word means? Here is another strategy. Say you've forgotten what 'interspecific' means. Think of an 'intercity' train, it goes between one city and a different one. Now you can work out what the term you've forgotten means. Don't put up a mental block to new or unfamiliar terms or ones you've just forgotten but use your existing knowledge base to help. You will be amazed how much you really do know!

REVISION PLANNER

The need to devise a strategy for regular and methodical revision can hardly be overstated. Last minute efforts for tests and examinations never work and usually result in panic and stress which are recipes for poor performance and underachievement. Leaving things too late nearly always means that any remedial help is at best fragmented and is insufficient to put right lack of knowledge and/or confidence. One of the best ways of avoiding this is to produce a planner of some kind. How this is constructed is highly individual but the sooner you work on this the better. It is important to tailor this to your own preferences but also to realistic goals. Try to break the course up into discrete manageable blocks and have a mechanism for flagging up areas which need revisiting as well as those which are not areas of concern. Use a consistent way of highlighting topics which need additional practice and/or help from your teacher. A great deal of self-discipline here will reap huge rewards as you move towards your final preparations for the National Examination.

Areas of Difficulty

Knowledge and understanding

Problem-solving

KNOWLEDGE AND UNDERSTANDING

Each year, the Principal Assessor for the SQA produces a list of areas where candidates have experienced difficulty in answering questions. Here is a collection of these for KU over a period of about six years, listed under their respective unit titles and subtitles:

Unit 1 Cell Biology

Cell Structure in Relation to Function

- difference between structure and function in cells
- difference between hypotonic and hypertonic
- osmosis and active transport
- role of respiration in generating ATP for active uptake of ions

Photosynthesis

- role of ATP and hydrogen in photosynthesis
- explanation of changes in GP and RuBP levels in chloroplasts under changing environmental conditions
- separation of photosynthetic pigments

Energy Release

- importance of ATP as a mean of transferring chemical energy in cells
- effect on cell growth of ATP produced in aerobic or anaerobic conditions

Synthesis and Release of Proteins

- importance of DNA replication for dividing cells
- role of mRNA
- detailed function of tRNA molecules during translation

Cellular Response in Defence in Animals and Plants

- detailed role of lysosomes in cellular defence
- role of lymphocytes in recognition of foreign antigens leading to tissue rejection
- phagocytosis
- stages in viral multiplication

Unit 2 Genetics and Adaptation

Variation

- characteristics of homologous chromosomes
- behaviour of chromosomes during meiosis
- relationship between genotype and phenotype

Mutation

- differences between gene and chromosome mutation
- effect of gene mutation on metabolic pathways/on amino acid sequences
- understanding the effect of mutations and natural selection in changing gene pools

Selection and Speciation

- concepts underlying speciation/importance of isolation
- adaptive radiation

Animal and Plant Adaptations

- relation of structure to function in root hair cells
- xerophytic adaptations
- difference between intraspecific and interspecific competition
- benefits of social defence mechanisms
- importance of net energy gain in the economics of foraging behaviour
- grazing and obtaining food

Unit 3 Control and Regulation

Control of Growth and Development

- sequence of events during induction of β-galactosidase
- role of α–amylase in germination
- role of IAA in apical dominance
- explaining locust growth pattern
- mechanism of phototropism
- definition of 'long-day plant'
- sun and shade plants
- pituitary hormones and photoperiod
- importance of nitrate for growth
- effect of light on flowering in plants and on the timing of breeding in animals

Physiological Homeostasis

- significance of vasoconstriction in low environmental temperatures
- significance of heat loss from mammalian body and the resultant increase in metabolic rate
- principle of negative feedback
- effect of ADH on kidney tubules
- importance of monitoring wild populations
- concept of population density

PROBLEM-SOLVING

Similarly, here are some areas of difficulty identified over the same period for data-handling type questions. These don't apply to specific areas of the course but are general, applying in different settings.

- calculations involving several steps
- calculations involving percentage changes
- calculations involving ratios
- establishing patterns from given data
- using data to justify an answer
- explaining a change

- explaining observations
- linking change to an effect
- failing to use the correct terms from a table heading in an answer
- evaluating experimental design
- purpose of precautions taken during an experiment
- identifying variables
- controls
- reliability
- drawing conclusions
- making predictions
- suggesting experimental design improvements
- accuracy in labelling graphs, plotting points and drawing a line graph
- scaling graphs to include 0 value

This book will go through examples which make particular reference to all of these areas known to cause problems. You will see what is needed to produce good answers and full marks, as well as examples of poor answers and an explanation of why those answers fell short of what was required. The examples will appear in the same setting as was raised by the Principal Assessor, that is as a multiple-choice, structured or extended-response type question. In every case, the original SQA questions have been used.

There will be plenty of exemplars of data-handling type questions to give you a clear understanding of how to overcome these difficulties. These will usually appear with KU questions or, sometimes, in isolation, just as they have done in previous National Examinations.

1 Multiple-Choice Questions

General advice

Examples

GENERAL ADVICE

In many ways, this is the easiest section of the examination because you don't have to recall the answers, merely recognise them. Each question is framed in exactly the same way: a stem followed by four answers, only one of which is correct. What varies is whether the question is KU or PS, the level of difficulty and the style of presentation. Look at these two examples on the structure of DNA:

1 One possible base-pair found in DNA is:

A adenine and thymine
B uracil and cytosine
C cytosine and adenine
D guanine and uracil

This is a simple KU question requiring a basic knowledge of the base-pairing in a DNA molecule and recognition of the correct answer.

2 The following set of results show the % base composition for a small mammal:

X	Y	Z	thymine
28.5	21.2	21.4	28.4

Which of these is a possible identification of the unknown bases?

	X	Y	Z
A	cytosine	thymine	guanine
B	guanine	adenine	cytosine
C	cytosine	thymine	guanine
D	adenine	cytosine	guanine

Essentially, this question is testing the same information but is much more difficult. Notice the format is not the same: the use of tables and figures makes this more like a PS question. First, you have to recall the complementary base-pairs since they are not given and then see that the percentage of thymine must equal that of adenine (here shown as X). The remaining two, Y and Z, must therefore be cytosine and guanine, either way round.

Providing you have a good knowledge base, you should be able to score well in this paper. The multiple choice paper is worth 30 marks, about 20 of which are for KU and 10 for PS. As a very rough guide, don't spend more than 1 minute on each question. In fact, if you know the answer, you can complete a question in a matter of seconds, releasing time for you to spend on other questions or parts of the paper which are not so easy. In principle, you will often know the answer before you see it below the stem. However, on those occasions on which you don't immediately know the answer, you may be able to eliminate two or even three of the options, allowing you to work out by this simple process what the correct answer is. Look at this example:

Which of the following organelles would you expect to find in particularly high numbers in an active sperm cell?

A ribosomes

B mitochondria

C lysosomes

D chloroplasts

You might not know at first glance what the correct response is but you might work along the following lines. Chloroplasts are not found in animal cells so D cannot be correct. The stem contains a cue word 'active' which suggests that sperm cells must require energy. Ribosomes are associated with protein synthesis not energy release, which eliminates A. Releasing energy in cells is associated with respiration, which in turn is linked to mitochondria, so B is very likely to be correct, if only based on an educated guess between the two options left.

Answer every question in this section; don't leave any unattempted. Wrong answers are not penalised here so, at worst, guess from the options left after you eliminate any which definitely can't be correct. Feedback from the SQA suggests students usually score about 20 out of 30 in this section, so you can check for yourself if you are hitting this as you practise. Of course, your aim is to score 30 out of 30 but, realistically, you may not be able to achieve that goal. Of the two areas, KU and PS, as you might expect, KU is usually completed best, especially in processing information and making conclusions.

Now let's work through some typical examples of multiple-choice questions taken from each of the three units. Try to see the general techniques which can be used to help you arrive at the correct answer.

EXAMPLES

Unit 1 Cell Biology

1 Which of the following proteins has a fibrous structure?

A Pepsin

B Amylase

C Insulim

D Collagen

You need to remember that enzymes are made of globular protein and since pepsin and amylase are both enzymes, A and B cannot be correct. Insulin is a hormone and is also a globular protein, making C incorrect. Collagen is a strong and inelastic material made of fibrous protein. This is a very basic KU type question typical of C grade.

2 Insulin synthesised in a pancreatic cell is secreted from the cell. Its route from synthesis to secretion includes

A Golgi apparatus → endoplasmic reticulum → ribosome

B ribosome → Golgi apparatus → endoplasmic reticulum

C endoplasmic reticulum → ribosome → Golgi apparatus

D ribosome → endoplasmic reticulum → Golgi apparatus

This is a more difficult type of KU question, requiring you to know the function of three different structures in the cell so you can correctly arrange the sequence of events to produce insulin. Like many hormones, insulin is a modified protein. Remember that ribosomes are associated with synthesising proteins so this stage of manufacture will occur first. That means A and C must be wrong. The endoplasmic reticulum essentially acts as a pathway or series of channels for transporting materials, such as proteins. The Golgi apparatus processes protein molecules before packaging them into membrane-bound sacs called vesicles for secretion. The correct sequence must therefore be D.

3 The action spectrum in photosynthesis is a measure of the ability of photosynthetic pigments to

A absorb red and blue light

B absorb light of different intensities

C carry out photolysis

D use light of different wavelengths for synthesis

The action spectrum is a graphical representation showing how the rate of photosynthesis is affected by light of different wavelengths. Photosynthesis occurs at the highest rate in the blue and red parts of the spectrum. A different graphical representation is used to show the absorption of light of different wavelengths. Taking this information into account, this is a simple KU question with D being the correct answer.

4 If 10% of the bases in a molecule of DNA are adenine, what is the ratio of adenine to guanine in the same molecule?

A 1:1

B 1:2

C 1:3

D 1:4

This is a simple PS question which requires you to be aware that adenine pairs with thymine and cytosine with guanine. The percentage thymine must be the same as adenine, i.e. 10%. This leaves 80% equally distributed between cytosine and guanine, each having 40%. The ratio of cytosine to guanine must therefore be 10 to 40, which is 1 to 4, making D correct here.

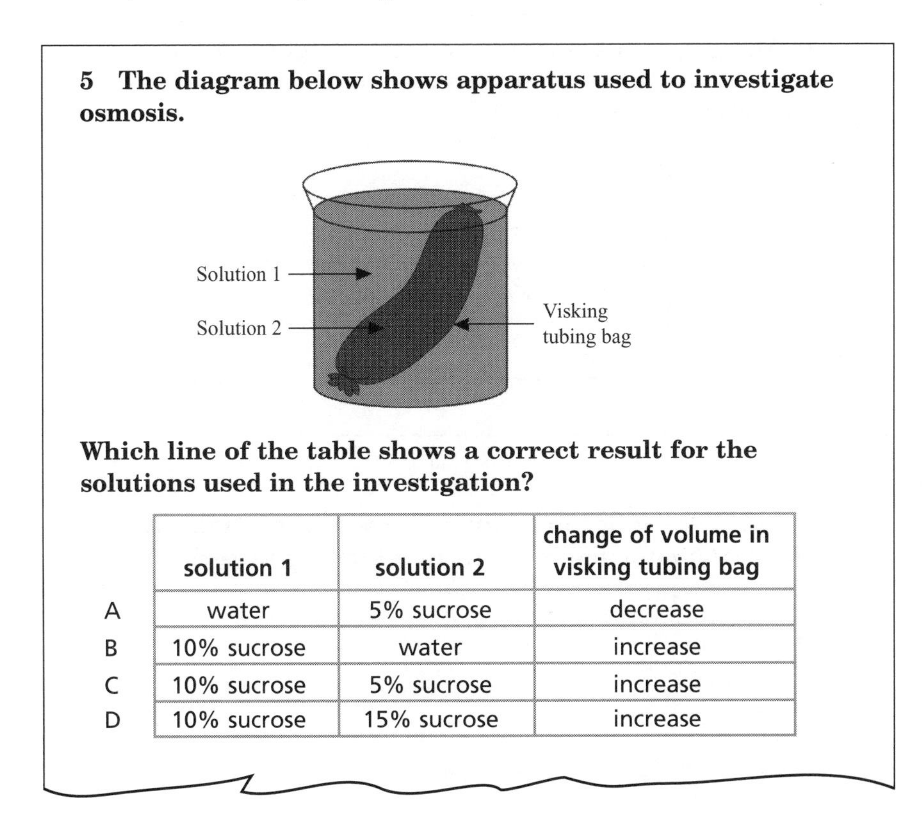

5 The diagram below shows apparatus used to investigate osmosis.

Which line of the table shows a correct result for the solutions used in the investigation?

	solution 1	solution 2	change of volume in visking tubing bag
A	water	5% sucrose	decrease
B	10% sucrose	water	increase
C	10% sucrose	5% sucrose	increase
D	10% sucrose	15% sucrose	increase

A more complicated PS type question here. First you need to understand what osmosis is, i.e. a special case of diffusion of water from an area of high to low concentration across a semi-permeable membrane. Once you have a clear understanding of this, the question is easier to understand. Notice, this is not a recall or recognition type question. You have to be able to use the information to work out the answer. First, consider in each case where the water concentration is highest and remember that the diffusion will take place from there to the other solution. In A, the water is highest in solution 1 so it will move from there into solution 2, but that will increase the volume of the bag, so this answer can't be correct. In C, the water is higher in 2 than 1 (since it is more dilute) so the water will move from 2 to 1, decreasing the volume of the bag, so C is wrong. In B, the water will move from 2 to 1, decreasing the volume of the bag, so this is wrong. In D, the water is higher in 1 than 2, so water moves from 1 to 2, increasing the volume of the bag, so this is correct. Notice that you must have this knowledge clear in your mind to start with before you can handle this type of question. Osmosis often causes problems, so it is worthwhile going over many questions to be sure you can handle them.

6 The graph illustrates the effects of light intensity, temperature and carbon dioxide (CO_2) concentration on the rate of photosynthesis.

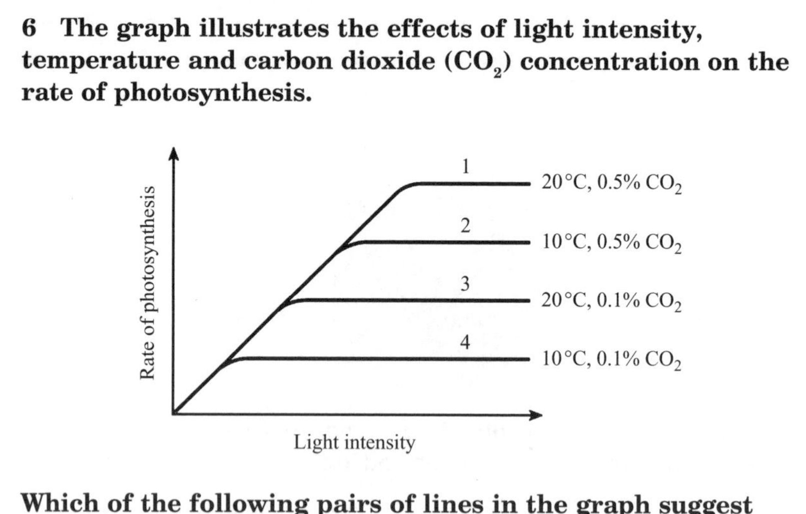

Which of the following pairs of lines in the graph suggest that carbon dioxide is acting as a limiting factor?

A 2 and 4

B 2 and 3

C 1 and 4

D 1 and 2

This is at the higher end of the PS type questions and one which only a well-prepared candidate would get right. It touches on a topic which many students have difficulty with, that of limiting factors. This is something which pervades biological systems, affected as they are by a wide variety of factors such as pH, temperature, light intensity, enzyme concentrations and so forth.

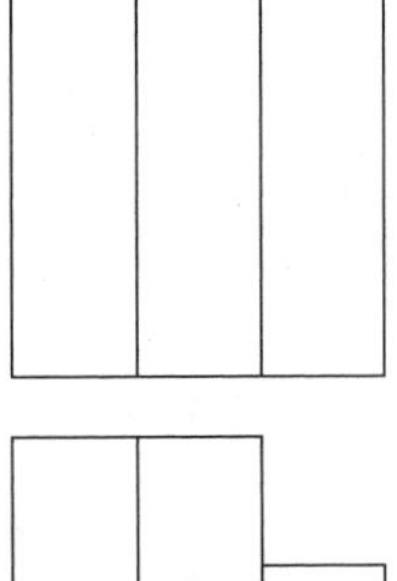

Here is an analogy to help you understand this concept before we tackle this question. Look at this diagram and imagine it is the face of a square box into which you can pour sand. The box is made of slats of wood. The volume of sand which the box can accommodate is not limited by the height of any one of the three slats.

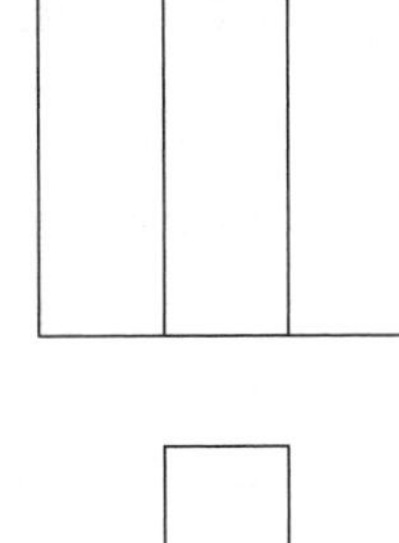

Suppose one of these slats is shortened, as shown. Now the volume of sand which the box can hold is limited by the height of the shortened slat, the limiting factor.

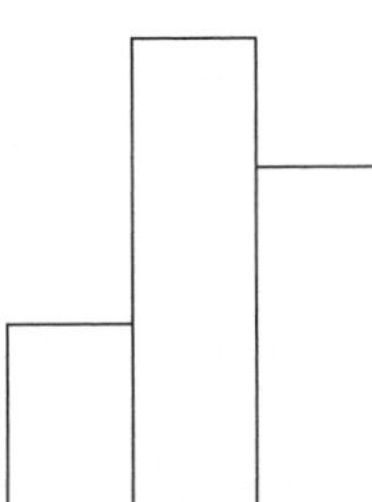

Next we shorten another slat, as shown. Notice the effect? Now the volume of sand which the box can hold is determined by the shortest slat, which has become the new limiting factor.

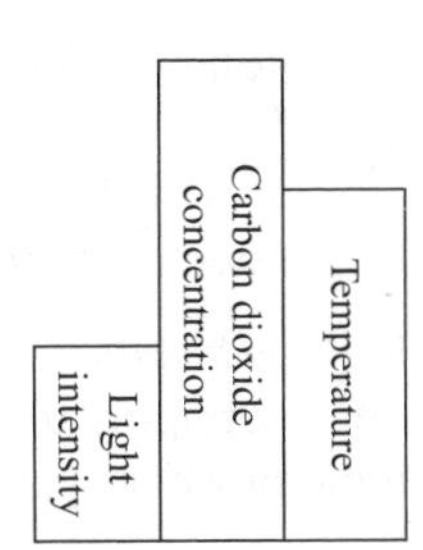

In a similar way, biological systems are affected profoundly by limiting factors. Indeed, we might redraw the last diagram like this for a system involving photosynthetic activity:

Notice that the limiting factor in this analogy would be light intensity, since this is the one which impacts first on the system, next would come temperature, and only then carbon dioxide concentration.

Now, with this understanding, let us consider the actual question which relates to CO_2 as a limiting factor. In order to establish which lines show this, we have to 'eliminate' the effect of the other factors, light intensity and temperature. With increasing light intensity the rate of photosynthesis increases, but eventually a point is reached where further increase produces no rate increase. At this point, other factors will be limiting. Comparing lines 1 and 2 with 3 and 4, which each have equal concentrations of CO_2 (0·5% and 0·1% respectively), shows that increasing temperature is the limiting factor. Only by comparing two lines with different CO_2 concentrations but the same temperature can we see the effect of increasing the CO_2 concentration. That would be either 1 with 3 or 2 with 4. The correct answer is therefore A.

EXAMPLES

Unit 2 Genetics and Adaptation

1 Which of the following descriptions does not relate to the process of meiosis?

A **Diploid cells are formed**

B **Two successive divisions occur**

C **Chiasmata often appear**

D **Homologous chromosomes appear**

This is a relatively easy KU question but it needs you to have a good understanding of the process of meiosis and how it differs from mitosis. As with other similar questions, three of these responses are correct and one is not. Let's go through them.

Meiosis involves two consecutive cell divisions so B is true.

Chiasmata, resulting in crossing over, occurs in meiosis so C is true.

Homologous chromosomes do appear in meiosis so D is correct.

Meiosis results in cells with half the normal (diploid) chromosome number so A does not apply to meiosis only mitosis. Notice here the link to Standard Grade Biology where you met the terms homologous chromosome and diploid.

2 Which of the following is true of the kidneys of a salt-water bony fish?

A They have few large glomeruli

B They have few small glomeruli

C They have many large glomeruli

D They have many small glomeruli

This is a slightly more difficult KU type question on water balance in a salt-water bony fish. These animals face the problem of losing water constantly to their salty environment by osmosis across the gills and parts of the gut. To balance this, they have to 'drink' the sea water and pump out the salt via their secretory cells but also pass out as little water in their urine as possible. The fish achieve this by having very few and small glomeruli, where ultrafiltration takes place, so that the filtration rate from the blood is low. The correct answer must therefore be B.

3 Which of the following statements regarding polyploidy is correct?

A It is more common in animals than plants

B It is the term used to describe the four haploid cells formed at the end of meiosis

C It results from crossing over of genes at chiasmata

D It can result from the non-disjunction of chromosomes

This is another straightforward KU question narrowed on one topic, that of polyploidy. Providing you have a good understanding of polyploidy, i.e. the presence of extra sets of chromosomes, you will have no difficulty with this question.

Polyploidy occurs rarely in animals because it actually interferes with the mechanism of determining sex, so A is wrong.

By its very name, the cells formed cannot be haploid, so B is wrong.

During chiasmata, sections of chromosomes exchange normally, not individual genes, so C is wrong.

Non-disjunction, the non-separation of chromosomes during meiosis, can result in gametes with more or less than the normal number of chromosomes.

Such abnormal gametes which carry more chromosomes than they should on fertilisation, will produce a cell with one or more extra sets of chromosomes (a polyploid in other words), so D is correct.

4 The table below shows the recombination frequency between genes on a chromosome.

Crossing over between genes	Recombination frequency
F and G	4%
F and J	6%
G and H	6%
H and J	4%

Use the information in the table to work out the order of genes on the chromosome.

The order of the genes is

A H G F J
B F G H J
C F G J H
D G H F J

This is a difficult PS question for many students on the position of alleles of genes along the length of a chromosome. The higher the frequency of recombination, the further apart the alleles will be. In order to work out the exact sequence, all the frequencies must be able to be placed on a straight line to correspond with the relative distances of the alleles compared to each other. Start with the first pairing, F and G, like this:

F - - - - G

Notice the use of 4 dashes to represent 4%. Next add another pair linked to one of these alleles. Let's chose F and J. Since F is 6 'units' away from J there are two possibles as follows:

F - - - - G - - J or J - - - - - - F - - - - G

Next we add G and H and again there are two possibles:

F - - - - G - - J - - - - H or H - - F - - - - G - - J

Notice at this stage either of these 'maps' is possible but when we have to add in the final pairing for H and J (4%), you can see only one map will work:

F - - - - G - - J - - - - H

The correct answer must therefore be C.

5 In guinea pigs, black fur (B) is dominant to white fur (b) and rough coat (R) is dominant to smooth fur (r).

Two heterozygous individuals are crossed and the possible genotypes of the offspring can be found using the Punnett square shown below.

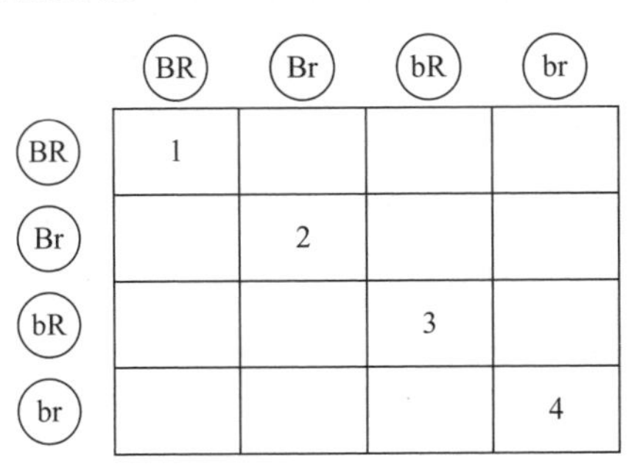

	BR	Br	bR	br
BR	1			
Br		2		
bR			3	
br				4

What are the genotypes of the individuals in the numbered boxes?

A Heterozygous for both gene pairs

B Homozygous for coat colour only

C Heterozygous for coat colour only

D Homozygous for both gene pairs

Students often find solving genetics problems hard but, with practice, this should not be the case. In fact, these questions all follow a similar pattern and you will almost always know if you are correct. Only by doing many examples and having them checked for errors and learning from these will you be able to master the techniques required. What you could do for this question is actually write in the genotypes onto to the Punnett square so that box 1 would be BBRR, 2 BBrr, 3 bbRR and 4 bbrr. Notice that in every one of these four cases, there is no example of heterozygosity, Bb or Rr, so A and C cannot be correct. That leaves B and D. In all four boxes, the genotypes for rough/smooth fur is homozygous, either dominant (RR) or recessive (rr) so B cannot be correct. In all four boxes, the gene pairs are each homozygous, so D is correct.

6 The base sequence of a short piece of DNA is shown below.

A--------G--------C-------T--------T--------A--------C--------G

During replication, an inversion mutation occurred on the complementary strand synthesised on this piece of DNA.

Which of the following is the mutated complementary strand?

A T--------C--------G-------A--------A--------T--------G--------A

B A--------G--------C-------T--------T--------A--------G--------C

C T--------C--------G-------A--------A--------T--------C--------G

D T--------C--------G-------A--------A--------T--------G--------C

This is a relatively difficult PS question because not only do you need to know what an inversion mutation is, you also must have read the question carefully to pick up that the answer is related to the complementary strand, not the original one given. Students often don't pick up on this and lose marks here. First, the complementary strand to the original without any mutation would look like this:

T--------C--------G-------A--------A--------T--------G--------C

An inversion causes a switching round of a pair of alleles on the chromosome. If the first pair had been inverted, the new mutated DNA would look like this:

C--------T--------G-------A--------A--------T--------G--------C

You can see only one sequence shows a possible inversion and that is strand C where the last two alleles have been switched round or inverted.

EXAMPLES

Unit 3 Control and Regulation

1 Which of the following is *not* the result of a magnesium deficiency in flowering plants?

A Curling of the leaves

B Yellowing of the leaves

C Reduction in shoot growth

D Reduction in root growth

Here is a slightly tricky question on the effect of a chemical, such as magnesium, on the growth of plants. Also, you are asked which one is NOT a result so you need to know all the effects of magnesium deficiency. A basic way to show this is to remove magnesium from the nutrient supply and see what symptoms the plant develops. Magnesium is known as a 'macro-element' because it is needed in relatively high quantities, especially for the manufacture of chlorophyll. A lack of magnesium will cause a lack of chlorophyll and thus a decrease in photosynthetic efficiency. Leaves will not look so green; instead they will yellow, so B would result. Growth is linked to photosynthesis so shoots and roots would be stunted, so C and D would result. Curling of leaves would not happen as a direct result of magnesium deficiency, making A the correct answer here. Notice how you can eliminate B, C and D to arrive at A.

2 The function of apical meristems in plants is to produce

A xylem vessels

B root hairs

C sieve tubes

D new cells

This is another very straightforward KU question which has elements of Standard Grade in it. Meristems are areas of plant tissue which have actively dividing cells, usually producing plant cells which have the ability to become specialised in their structure and function. Since xylem vessels, root hairs and sieve tubes are all specialised already, answers A, B and C cannot be correct. D is clearly the answer here.

3 The Jacob-Monod model of gene expression involves the following steps.

1 Gene expression

2 Exposure to inducer substances

3 Removal of inhibition

4 Binding to repressor substance

The correct order of these steps is

A 2, 4, 3, 1

B 3, 4, 2, 1

C 4, 1, 2, 3

D 1, 4, 2, 3

You need to have a good understanding of the model proposed by Jacob and Monod in 1961 to be able to work out which answer is correct. Diagrams of this model would be a very effective way of making sure you understand the interplay between the different elements of the model.

An inducer substance, such as lactose, combines with the repressor substance thereby removing the inhibition of the operator gene. This gene is now switched on and therefore allows the expression of the gene(s) to transcribe the protein which might be an enzyme such as β-galactosidase. The sequence must therefore be 2, 4, 3 then 1 so A is correct.

4 If the body temperature drops below normal, which of the following would result?

A Vasodilation of skin capillaries

B Vasoconstriction of skin capillaries

C Decreased metabolic rate

D Increased sweating

This is a relatively easy KU question on temperature control. Humans maintain their body temperature within very narrow limits and invoke various mechanisms to achieve this steady state. Such mechanisms include shivering to generate heat from involuntary muscular contractions, sweating to cool the body as the liquid evaporates from the skin, changing the diameter of the tiniest blood vessels, called capillaries, near the skin surface and altering the rate of metabolism. The more blood which passes near the skin surface, the more heat is radiated away. Metabolism is a collective name for the biochemical reactions of the body which will generate heat as these are carried out. You also need to understand the terms vasodilation and vasoconstriction which mean a widening or narrowing of blood vessels respectively. If the environmental temperature falls below normal, causing the body's temperature to start to drop, automatic responses are triggered into action to try to compensate for this. For example, sweating is reduced to prevent radiative loss by evaporation, making answer D wrong. Metabolic rate rises to make more energy available as heat, making C wrong. Blood will be diverted away from the skin surface to prevent radiative loss, making A wrong, leaving B as the correct answer since the vessels will indeed narrow when the temperature falls below normal.

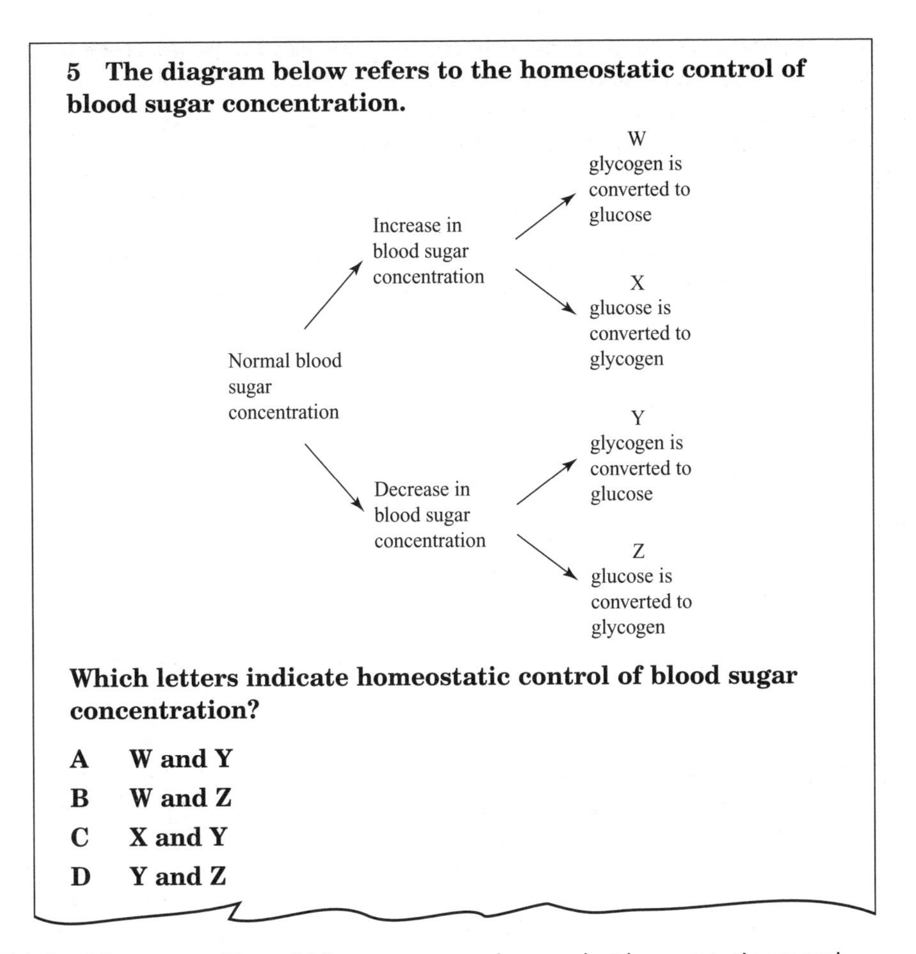

5 The diagram below refers to the homeostatic control of blood sugar concentration.

Which letters indicate homeostatic control of blood sugar concentration?

A W and Y

B W and Z

C X and Y

D Y and Z

This is a PS type question which assumes your know what homeostatic control means. Like many body variables, blood glucose concentration is very carefully monitored and controlled in the body to keep it at a constant level. That control is called homeostasis but remember it applies much more widely to temperature, carbon dioxide levels and so on. For this question, you need to know which is the storage carbohydrate, i.e. glycogen, and which is the blood sugar, i.e. glucose, and how these are interchangeable according to demand. With a low blood sugar level, you need more glucose released and conversely, with a high blood sugar level, you need less.

Since W produces more glucose in the blood stream, it can hardly help if there is an increase in blood sugar concentration so A cannot be correct. Notice, we don't need to work out Y here.

In a similar way, B must be wrong.

Converting glucose to glycogen will happen when blood sugar is increased so it applies in both C and D. Notice now we have to work out Y and Z.

Decreasing blood sugar level will not trigger the conversion of glucose into the insoluble glycogen so D is wrong.

Decreasing blood sugar level will raise the demand for glycogen to be converted to glucose, Y. With X, now we have two reactions opposite in effect to control blood sugar level, making C correct.

6 The graphs below contain information about the population of Britain.

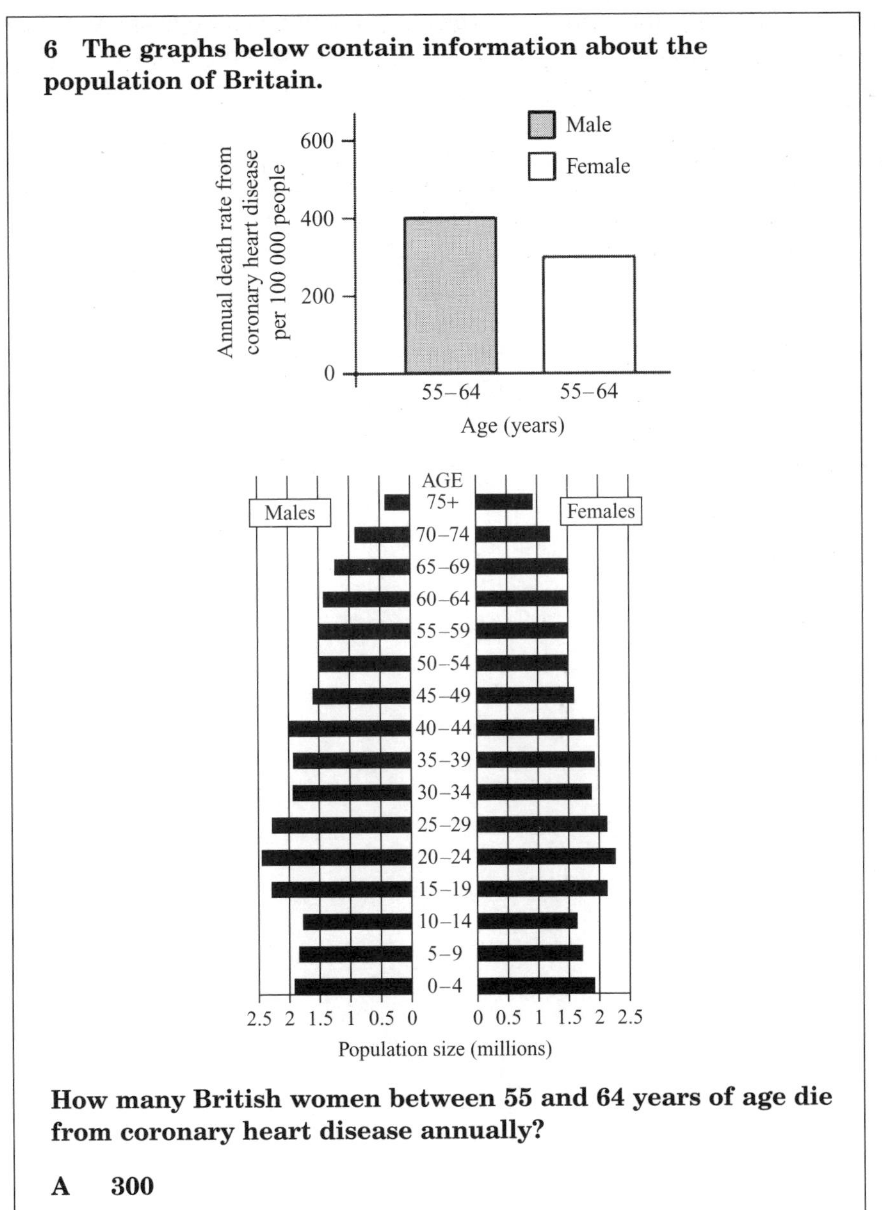

How many British women between 55 and 64 years of age die from coronary heart disease annually?

A 300

B 4500

C 9000

D 21 000

This is a difficult PS question testing the area of monitoring populations. Do not be put off by the quantity of data shown here in the graphs. Take your time and read the question carefully, perhaps a few times, until you understand what is being shown here and what is being asked.

Since the question relates to women, the data for the males is not relevant to this question. You need to use the top graph to get the annual death rate from coronary heart disease per 100 000 people, which is 300 for women. Next, from the second graph, you need to read the size of the female population for the age groups 55–59 and 60–64 since you are asked for that range in the question. This is 1·5 million for both so the total is 3 000 000. Notice how the data is easy to read from the graph. Of the 3 000 000 there will be 300 deaths in every 100 000. By dividing 3 000 000 by 100 000 we get 30, so we now multiply this by 300 to get 9000, making answer C correct.

2 Structured Questions

General advice

Examples

GENERAL ADVICE

In this section, unlike the multiple-choice paper, you will not have answers provided. Instead you have to generate the answers from the information given and your own knowledge base. There are no choices here and you are expected to answer every question. Guesses are not penalised so, at worst, have an educated guess. Usually the stem of the question guides you as to the type of answer needed. For example, if you are asked 'What is the likely optimum temperature for this reaction?', a guess would need to take the form of a number plus the correct unit °C. It is surprising how many candidates don't complete every question even though they have time to go back and make a best guess.

You can afford to spend more than 80 minutes on this paper. Remember you have 150 minutes to answer 130 marks of questions in total and you will probably pick up time from the multiple-choice section, providing you handle the questions efficiently and effectively. However, do keep an eye on the clock and try not to spend more than 90 minutes here before moving on. Timing is vital to success, so it is a good idea to practise these questions against the clock so you develop a good self-awareness of how long they take to complete. Don't dwell too long on any one part but move on to the next question if you are not sure of an answer, trying to pick up marks for questions you feel confident about. You will almost certainly have time to come back to those questions which are incomplete. Putting a small, consistent mark against these helps flag them up quickly and easily for a second visit. Often your brain works on these subconsciously and, on a second read through, you suddenly find you do know the answer.

Of critical importance is what is called the weighting attached to questions in this section. As a general rule, each mark needs its own answer, so if 2 marks are allocated to a question, it will require two points to obtain full marks. Here again, many candidates slip up by not taking notice of this most basic advice. Look at these two examples, both from a SQA specimen paper:

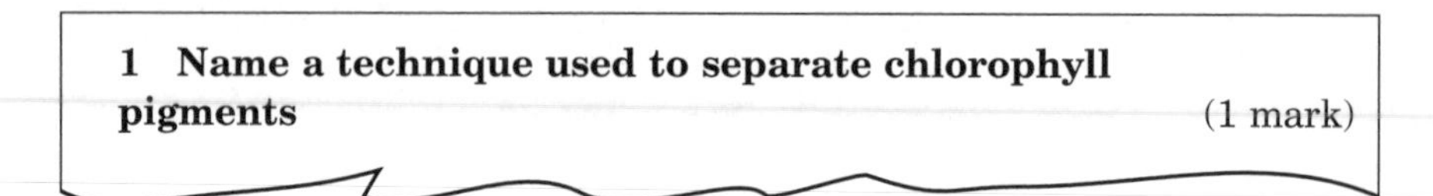

1 Name a technique used to separate chlorophyll pigments (1 mark)

Here a simple answer of chromatography is all that is needed.

2 Explain why protoplasts have to be kept in a salt solution of the same concentration as the cell sap and not in distilled water (2 marks)

Now two distinct points are needed to obtain full marks. First, the isotonic salt solution will prevent water entering by osmosis and secondly, since protoplasts have no cell wall, such osmosis would cause cell damage. Notice how full this answer is compared to the previous one. As another guide, look at the space you have for an answer. Is it one line or two or three? This helps guide you as well.

While the multiple-choice questions each tend to focus on one area of the course, questions in this section can spread across other parts of the course. For example, a question on active transport might connect to aerobic respiration and the supply of energy via adenosine triphosphate. This is why it is important not to compartmentalise your knowledge base but see connections between different parts of the course.

As a general rule, don't use abbreviations, though in some cases these will be accepted. For example ATP will normally suffice for adenosine triphosphate, as will RNA for ribonucleic acid and ADH for anti-diuretic hormone. However, if there is any doubt at all, use the full term, for example uracil instead of U.

While spelling should be correct, it is easy to make mistakes or even forget completely how to spell a term correctly. The examination authority will make every effort to give you the benefit of the doubt so, for example, if you spell 'protein' as 'protien' or 'chlorophyll' as 'klorophyll' you will not be penalised. What you need to watch is where words overlap a little. A good example of this would be these words: ureter, urethra and uterus. You need to be very careful here, obviously.

Read the question carefully. It may seem obvious to say this but examiners will tell you that candidates often go off at a tangent to what has been asked because they do not understand what has been asked in the first place. Does the question ask you to 'state' or 'explain', for example? If you are asked to describe or explain something, a one-word answer will not get any marks here.

Sometimes a candidate may give more than one answer to a question. For example, suppose you were asked: 'From the graph, what would be the effect of increasing the temperature of this reaction from 20°C to 30°C?' Let's say you wrote: 'An increase from 20°C to 30°C would make the reaction speed up because the enzyme was denatured.' Notice you have gone beyond what was asked and, in doing so, have added something which is not only wrong but in fact almost contradicts, or negates, the first part of your answer. In cases where the second part of answer negates the first, you may not get the mark.

If the answer is in the form of a number and units are not given in the answer space, then you must include these in your answer if they appear in the stem of the question. This is one of the most common errors which even the very best candidates make, forgetting to include units such as g/secs/mins/m^2 /°C and so on. If you can devise a way of asking yourself each time, 'Have I checked for units?', this will very likely earn you extra marks somewhere. If you do forget the units for a question and there are follow-on parts within the same question which also require those units, you will not be penalised twice, in other words, the error is not carried.

In a data-handling question which is in parts, as they often are, if the second or subsequent part is correct but based on a previously given incorrect answer, you will not be penalised twice.

Where, for example, 2 marks are available and 4 answers are required, a sliding scale will apply. It could be that 4 correct answers will get 2 marks, 2 or 3 correct answers 1 mark and 1 correct answer will get 0 mark.

Make sure you use the data provided and don't go beyond what you are allowed. For example, suppose you had to use a key to identify or give features of viruses and you were asked to describe the virus causing acquired immune deficiency syndrome. Let's say the following information was in the key about this virus: retrovirus, glycoprotein markers on outside, nucleic acid present. If you were to write 'this virus has reverse transcriptase' but this was not given in the key, you would not be answering the question, even if that piece of information was technically correct. In other words, you are being tested in part on your ability to stick to the question and use the data provided.

This part of the paper also carries two special questions. One question, worth about 8 marks, will be on a piece of practical work which almost certainly you will not have carried out yourself. This question tests your ability to think rather than recall facts. For example, you might be asked: 'How you would make the experiment more reliable?' or 'How you would set up a suitable control?' In addition, there will be a question on data-handling and problem-solving skills also worth about 8 marks. This again will often be about a procedure or experiment which you will most likely not have met before. It might ask you to 'Draw a line-graph of these results', for example. You should not be concerned that these questions will present new practical contexts since all the information will be there for you to make predictions, analyse, present data in a new way, and so forth. With experience and practice, you will become familiar with the style of these questions.

Use a pencil when drawing figures and, if asked to label any structure, do so with lines but no arrows, ensuring that each line ends unambiguously on what is being labelled. When drawing line-graphs, histograms or bar-charts, use a ruler throughout. Join plots point to point with your ruler and don't do a best-fit or smooth curve. Ask yourself: 'Have I labelled each axis correctly? Have I used more than half the grid by choosing a suitable scale? Have I used the correct units?' Students often get confused here with the concept of independent and dependent variables and which goes where on a graph. The dependent variable is the one which you observe changing as you change the independent one, the one you have control over. Suppose you were measuring the change in pulse rate as someone was exercising over a period of time. The independent variable here is time, the variable you are controlling by measuring over some definite agreed time interval, say every 5 minutes. This is plotted on the x-axis. The dependent variable is the pulse rate, the one you observe. This is plotted on the y-axis. When you are reading off a graph, you will usually find the point is very clear so try to be as precise as you can be and not rush obtaining the value.

Sometimes the question will specify 'to the nearest whole number' in which case you must round up your answer, but otherwise don't answer to more than one or two decimal places.

Often, there are alternative answers which are acceptable. Some, but not all, of these are indicated in the text which follows by '/' so that '*ATP/energy source*' would mean either is allowed. The actual past paper books will give a full list of these.

In order to make it easier for you to use this section, the questions will not appear exactly as they do in the SQA examinations. In most cases, the answer lines will have been removed and the questions broken up by suggested answers and commentary.

EXAMPLES

Unit 1 Cell Biology

1 (a) Samples of carrot tissue were immersed in a hypotonic solution at two different temperatures for 5 hours. The mass of the tissue samples was measured every hour and the percentage change in mass calculated.

The results are shown on the graph.

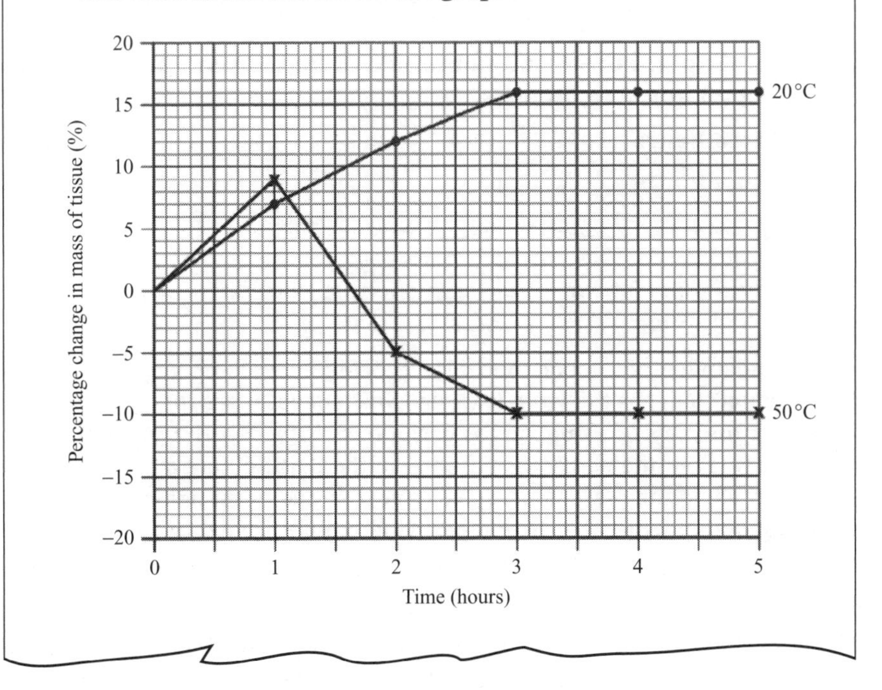

This question touches on several areas which cause problems for students: hypotonic and hypertonic solutions as well as osmosis and active transport. Let's briefly clarify the distinction between hypotonic and hypertonic solutions.

First, these terms are usually relative, that is you are comparing one solution with another. So to say a solution is hypotonic means it is more watery than another and, by contrast, a hypertonic solution is less watery than another. If two solutions have the same water content, they are said to be isotonic. Suppose you pour two cups of tea of equal volume. They will be isotonic to each other.

Now you add sugar to one, making that one hypertonic to the other, i.e. it is now less watery than the other. You could also say the other one was hypotonic to the one to which sugar was added, since it is more watery.

When a piece of plant tissue is placed in a hypotonic (watery) solution, water will tend to move in by osmosis so the mass will increase. Notice that it is common to express changes in mass as percentage changes because often the initial masses of the samples are different.

Let's go through the question part by part and see how different candidates might tackle each part. A weak candidate's attempt is shown first, followed by a good candidate's attempt which would obtain maximum marks.

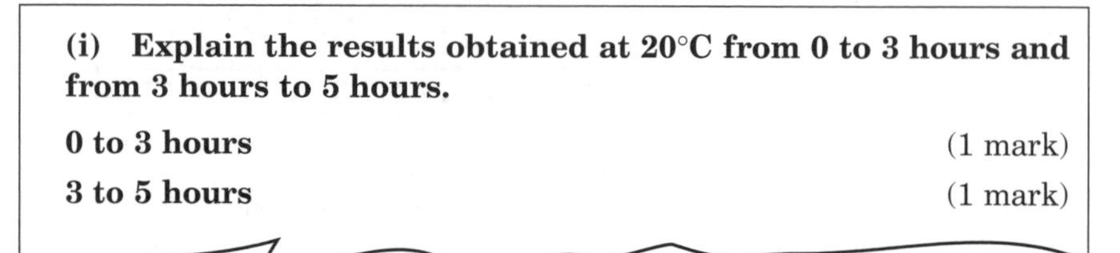

(i) Explain the results obtained at 20°C from 0 to 3 hours and from 3 hours to 5 hours.

0 to 3 hours (1 mark)

3 to 5 hours (1 mark)

(i) 0 to 3 hour *the mass of tissue increases*
3 to 5 hours *it stays the same*

This candidate has not understood what is being asked here. Very importantly, the question stem has the word 'explain' which needs more than just the statement *the mass of tissue increases*. Notice that the y-axis shows not the mass of tissue but the percentage change in the mass so here a basic error was made in reading the graph labels. The correct answer must demonstrate an understanding of why the percentage change climbs from 0 to 15% over these 3 hours. No marks awarded here.

Similarly *it stays the same* does not answer this question. Also, it demonstrates a very common error on the part of candidates by assuming the examiner will know what 'it' is referring to. Remember, an examiner is not allowed to interpret your answers, so make it very clear what you mean. This candidate has seen that the graph has levelled off between 3 and 5 hours but not said why. No marks awarded here.

Now, by contrast, here are the good candidate's answers.

(i) 0 to 3 hours *water is taken into the plant tissue by osmosis*
3 to 5 hours *no further water can enter because the cells are turgid*

Notice how short but very appropriate these two answers are, clearly demonstrating an understanding of what is happening and why the changes take place. In addition, this candidate has used the technical term 'turgid' to describe the plant cells when fully charged with water. You can keep answers short like this providing they are correct and relevant, as this one is. Other possible answers for 0 to 3 hours might be *water diffuses into the cell* or *water enters the cell from a more dilute solution* or *water moves down a concentration gradient.* For 3 to 5 hours, other possible answers could be *cells are isotonic with the solution* or *cell wall cannot expand any more.*

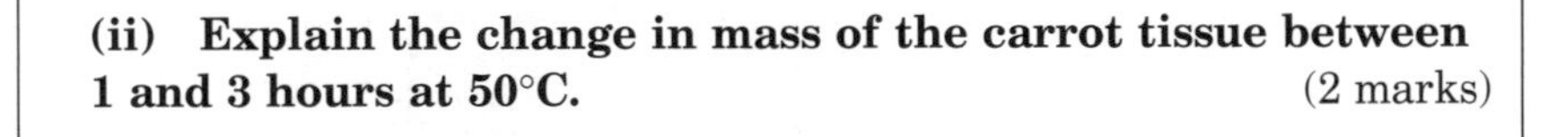

(ii) Explain the change in mass of the carrot tissue between 1 and 3 hours at 50°C. (2 marks)

(ii) *the carrot has been killed at 50°C so the mass of tissue falls*

Notice that this question carries 2 marks and has a relatively large space for completion. This tells you that one point, even if correct and relevant, will not earn both marks. This candidate has given only one point so immediately has lost the other. The answer given shows a common misunderstanding about membranes in cells, i.e. that they will be 'killed' by high temperatures. Membranes are collections of molecules and not alive in the conventional sense. High temperatures, such as 50°C, will certainly destroy them by denaturing protein and causing the membrane to lose its ability to regulate what enters or leaves the cell. This doesn't happen immediately; it would take time for the high temperature to penetrate the plant tissue and cause all the membranes to be destroyed, stopping osmosis taking place and then causing disintegration of the tissue itself. No marks would be awarded here.

(ii) *High temperatures cause the membrane proteins to be denatured. The cells therefore will no longer be semi-permeable and therefore the contents will leak out*

Two distinct points which link together are given. High temperature causing denaturation or destruction of the membrane proteins causing the loss of semi-permeability which in turn then allows the contents, such as water, to leave the cell.

This candidate has given two correct explanations which link together, the second following on from the first.

(b) The chart shows the concentration of ions within a unicellular organism and in the sea water surrounding it.

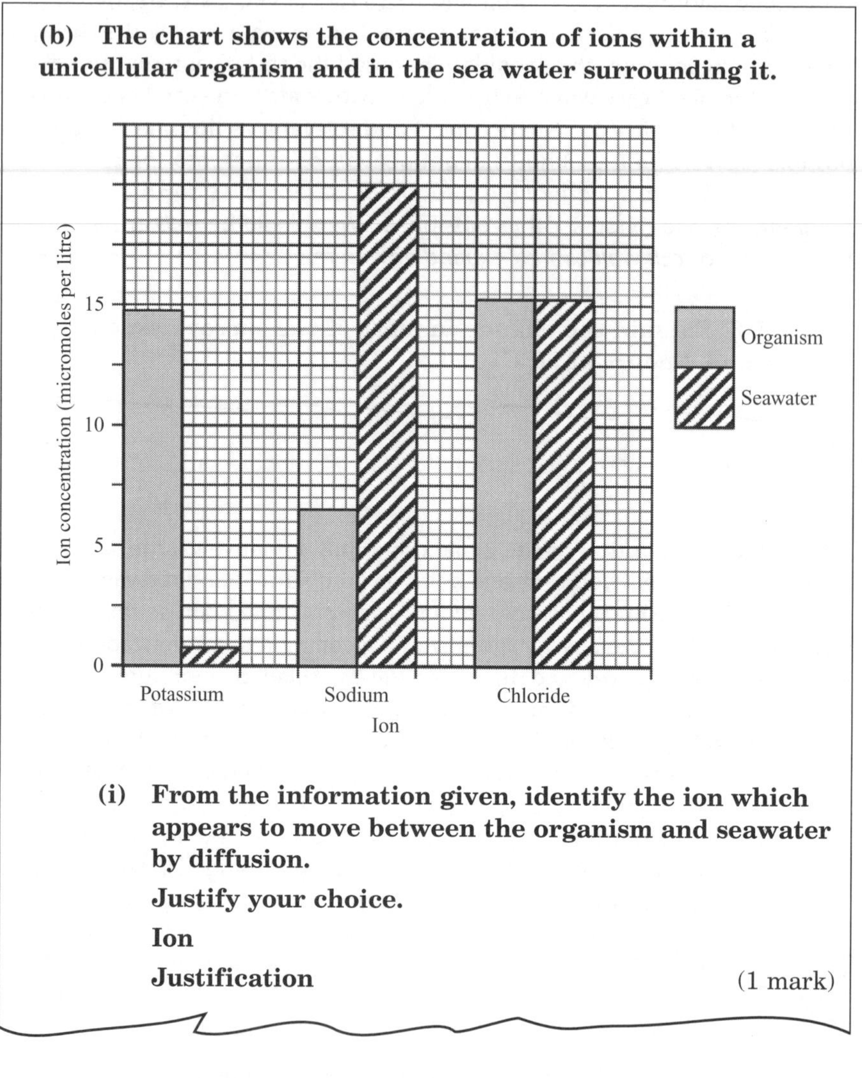

(i) From the information given, identify the ion which appears to move between the organism and seawater by diffusion.

Justify your choice.

Ion

Justification (1 mark)

(i) chloride
it's the same

First, notice the question asks 'identify the ion …' not 'identify an ion …' which tells you that only one answer is possible here. Try to pick up on that type of clue in questions in other settings. The ion is indeed chloride but the weak candidate has not justified the choice by bringing out the equal concentrations of the ion inside and outside the organism. No marks awarded here.

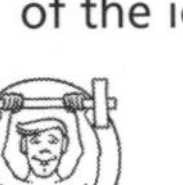

(i) *chloride*
the ion concentration outside and inside the organism is the same

Again, a short but clear justification.

> **(ii) When oxygen was bubbled through a tank of seawater containing these organisms, the potassium ion concentration within the organism increased.**
>
> **Explain this effect.** (2 marks)

(ii) *oxygen helps keep the organism alive so it can take in more potassium*

The candidate has indeed given two distinct points here but neither answers the question, which calls for an explanation. Notice how the answer given '... *can take in more potassium'* almost restates part of the original question: '.... the potassium ion concentration within the organism increased.' You cannot expect to obtain marks for this type of approach. Similarly, don't use words or terms which appear in one question to answer another. Papers are made up by experts who check carefully that no clues are given in this way so you will almost certainly not get any marks for this type of borrowing of terms. While oxygen will help keep the organism alive, this does not explain the link between increasing oxygen supply and increasing uptake of potassium ions. No marks awarded here.

(ii) *active transport of potassium ions is energy-demanding; this energy is obtained from aerobic respiration which is increased by increased oxygen availability*

A full answer here showing a clear understanding of active transport being energy-demanding which in turn needs aerobic respiration to take place. Aerobic respiration is highly dependent on the supply of oxygen to supply energy in the form of ATP.

2 ***Figure 1*** **shows how glycerate phosphate (GP) and ribulose bisphosphate (RuBP) are involved in the Calvin cycle.**

Figure 1

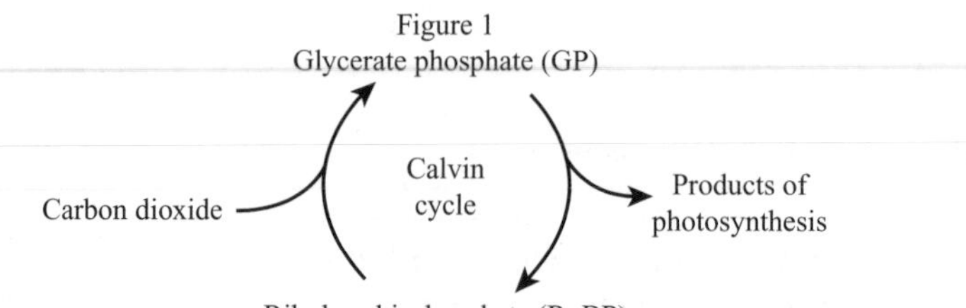

An investigation of the Calvin cycle was carried out in *Chlorella*, a unicellular alga.

Graph 1 **shows the concentrations of GP and RuBP in *Chlorella* cells kept in an illuminated flask at 15°C. The concentration of carbon dioxide in the flask was 0·05% for the first three minutes, then it was reduced to 0·005%.**

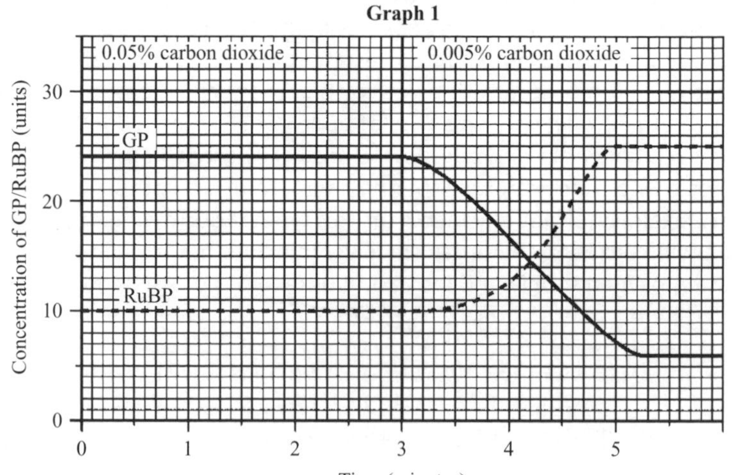

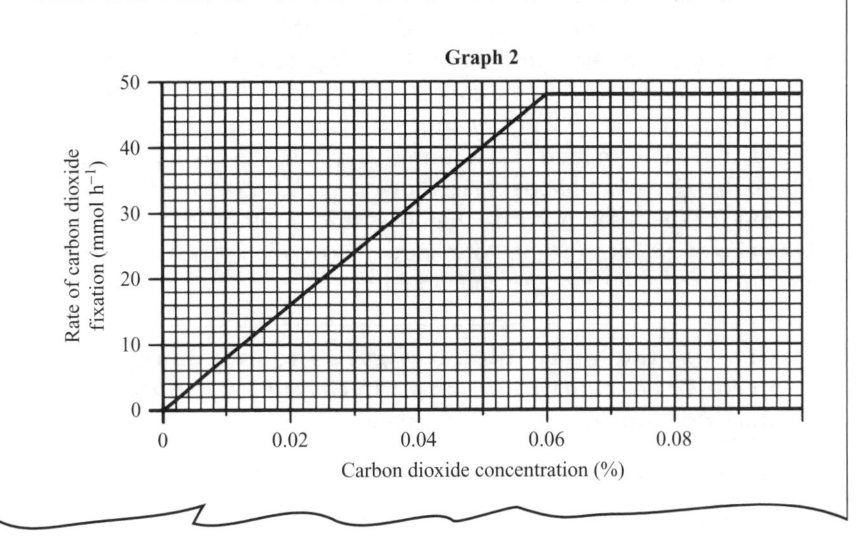

Students often find it tricky to understand how various chemicals involved in the photosynthetic process change when environmental conditions change. Like all biochemical processes, photosynthesis is affected by factors such as temperature, pH, light intensity and so forth. This question looks at the effect of changing the carbon dioxide concentration on the concentration of glycerate phosphate and ribulose bisphosphate as well as the rate of photosynthesis itself.

This question also tests your ability to read information from graphs carefully and then do some basic calculations.

We'll consider the answers given by a weak student and see how these can be improved on.

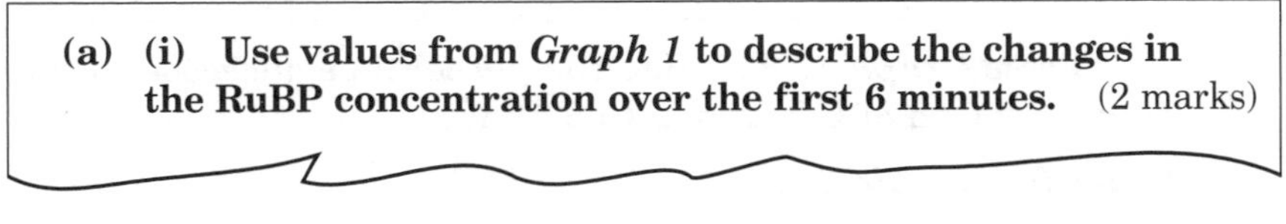

(i) *it stays the same then increases and levels out*

Notice this question leads you to a particular graph by using bold letters for '**Graph 1**' so make sure you always follow this when answering and don't use information from another part. The student starts off with *it* expecting the examiner to know what is being referred to, another common error. Make your answers explicit; in other words, make them clear to the examiner. Instead of *it* the student should have said *the concentration of RuBP* which is now clear. The question refers to 'changes' not 'change' so more than one has to be described. When you have information in the form of a graph, use the figures, not just words, to describe what happens. See how the concentration of RuBP starts off at 10 units for the first 3 minutes, then rises to 25 units over the next 2 minutes and remains at 25 units for the last minute. This student has not used any figures to describe the changes and would get no marks.

(ii) Use the information in *Figure 1* to explain the increase in RuBP concentration shown in *Graph 1* when the carbon dioxide concentration is decreased. (2 marks)

(ii) RuBP uses carbon dioxide to make GP which is changed back to GP

This is a difficult question needing you to bring together information presented in two different forms. Figure 1 shows a broad view of the Calvin cycle so that you can see RuBP uses up carbon dioxide to form GP which the student correctly observes. However, the student does not fully explain why, as carbon dioxide concentration is changed from 0·05% to 0·005%, the RuBP concentration rises as shown in Graph 1. The answer is almost correct, obtaining 1 mark, but needs to go a little further to get the second mark. A better answer might be: *'RuBP uses up carbon dioxide to form GP so a decrease in carbon dioxide concentration will mean less RuBP is converted. The GP being changed to RuBP is not affected so RuBP continues to be produced.'* This is a very full answer and would get 2 marks.

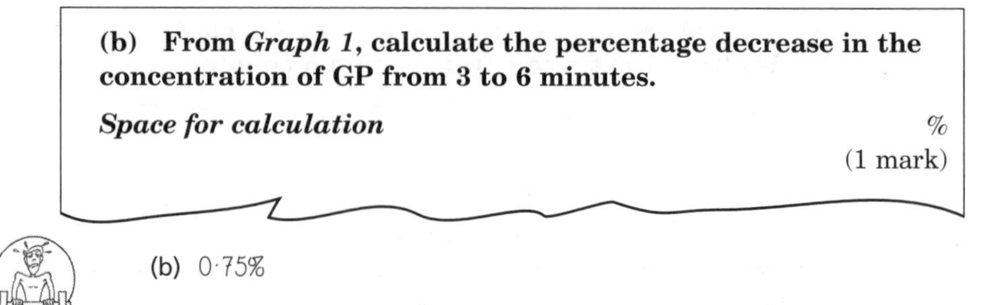

(b) From *Graph 1*, calculate the percentage decrease in the concentration of GP from 3 to 6 minutes.

Space for calculation %

(1 mark)

(b) 0·75%

This is a simple calculation which does not even require the use of a calculator. The figures from the graph have been made easy for you to work out the percentage change. This is usually calculated by the change divided by the original value then multiplied by 100. Let's see how this applies here.

The value for GP concentration over the first 3 minutes is 24 units. It fall to 6 units 3 minutes later, a decrease of 18 units. The calculation is therefore:

$$18/24 \times 100 = 3/4 \times 100 = 75\%$$

This student almost completed the calculation correctly but did not multiply by 100 so no marks would be awarded here.

(c) Use the terms 'increase', 'decrease' or 'stay the same' to complete the sentence below. Each term may be used *once*, *more than once* or *not at all*.

If the carbon dioxide concentration was returned to 0·05% at 6 minutes,

the concentration of RuBP would

and the concentration of GP would (1 mark)

(c) decrease
increase

This is a very straightforward question asking you to predict what will happen if the conditions of carbon dioxide concentration are restored to what they were at the start. Both answers are needed here to get the 1 mark as this student has done.

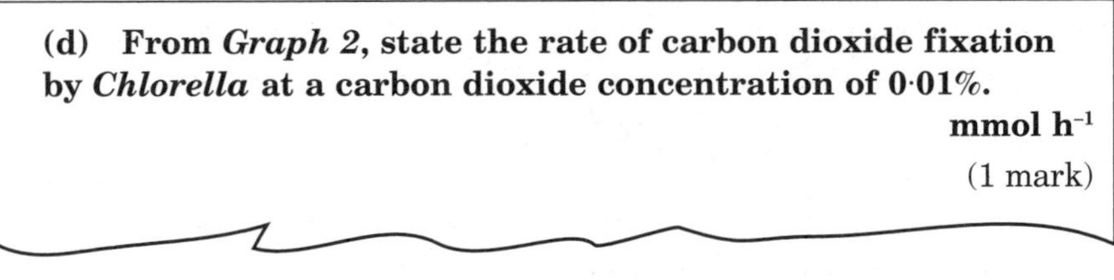

(d) From *Graph 2*, state the rate of carbon dioxide fixation by *Chlorella* at a carbon dioxide concentration of 0·01%.

mmol h^{-1}

(1 mark)

(d) 9 mmol h^{-1}

Here you need to read the y-scale carefully on Graph 2 and note that each box represents 2 units not 1 as in Graph 1. Don't assume the scales are the same between graphs. This student has made such an error and has, accordingly, given a wrong answer. The correct answer should be 8 mmol h^{-1} so no marks are awarded here.

(e) How many times greater is the rate of carbon dioxide fixation from 0 to 3 minutes compared with 3 to 6 minutes?

Space for calculation **times**

(1 mark)

(e) 20 times

This is the most difficult part of the whole question and really is aimed at A grade. From Graph 1, between 0–3 minutes the concentration of carbon dioxide was 0·05% and between 3 and 6 it was 0·005%. These translate into rate of carbon dioxide fixation of 40 and 4 mmol h^{-1} from graph 2, an increase of 10 times. Notice how this most difficult part comes last so that you have a chance to become familiar with the question in the easier sections first. No marks would be awarded here.

3 The diagram below shows a simple respirometer. This apparatus, with a suitable control, was used to compare the rates of respiration of various living materials at 20°C.

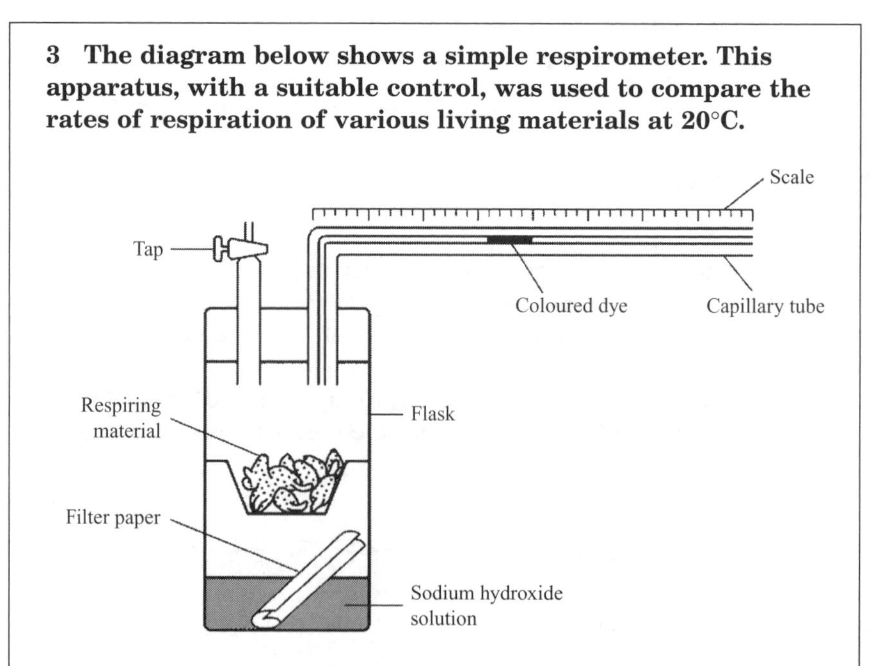

The living material was placed in the flask with the tap open. After 10 minutes the tap was closed. The apparatus was left on a bench in the light for 5 hours, during which the rate of respiration was measured.

The results obtained are shown in the table below.

Respiring material	Relative oxygen uptake (mm per hour)
Germinating seeds	*2*
Earthworms	*8*
Woodlice	*11*
Mealworms	*7*

In the National Examination, you will be expected to answer a question worth around 8 marks on an experimental procedure which almost certainly you have not carried out. This is designed to see if you can work things out yourself rather than simply recall facts. It will touch on experimental design, perhaps how it was flawed and might be improved, controls, precautions, etc. There may be a link to a graphical display of data, such as drawing a bar graph as here, as well as linked questions to particular areas of the course, here respiration. Do not be fazed in any way by the presentation of a new situation. Take your time and calmly go through the question section by section. Let's do that now.

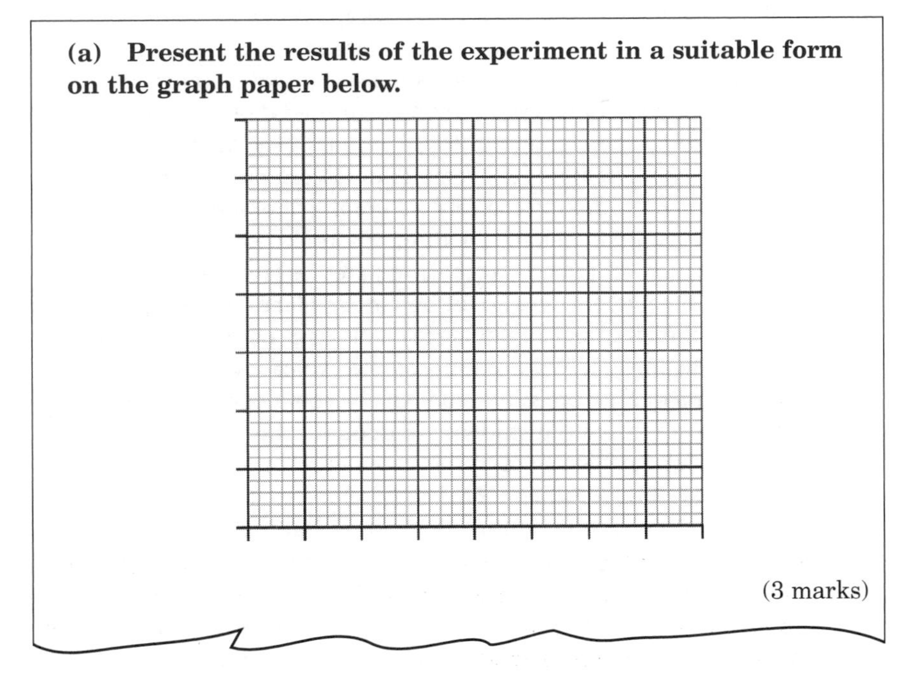

(a) Present the results of the experiment in a suitable form on the graph paper below.

(3 marks)

(a)

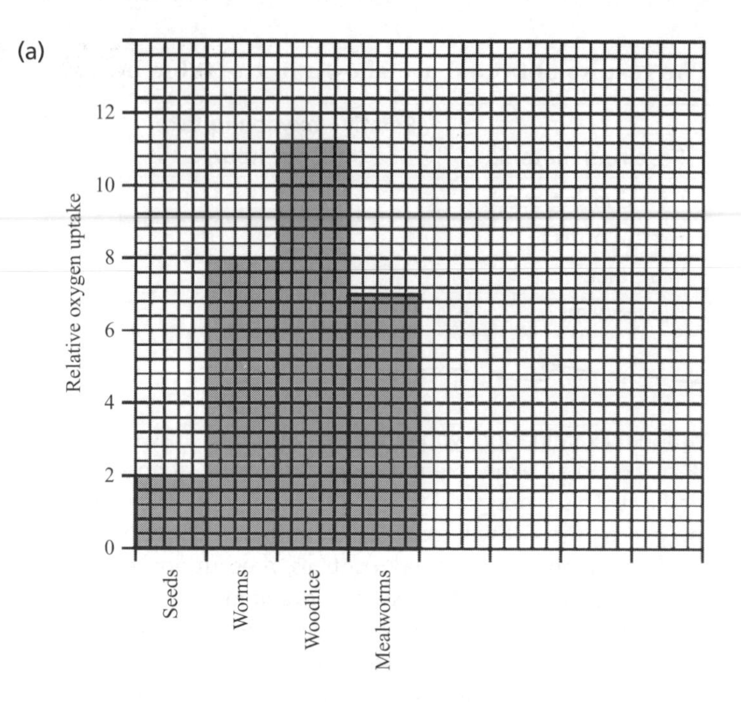

The first thing to ask yourself is what type of graph is needed here. Sometimes the question will actually specify this but not here so you need to work this out. Generally, when plotting variables which are discontinuous, you should use a bar chart. Discontinuous variables cannot be measured directly with a balance, stopwatch, ruler, thermometer and so forth. Blood group, right/left handedness and eye colour in humans are clear examples of discontinuous variations which can't be measured. Even though the relative uptake of oxygen can be measured, the variable of 'respiring material' cannot so this indicates a bar chart is needed. Next, the construction of a good bar chart requires careful drawing so that you obtain all the marks. Each axis must be properly labelled, using the precise naming given in the question. This student contracted 'germinating seeds' to 'seeds' and 'earthworms' to 'worms' for example, which is bad practice. Units must be included and their omission is one of the commonest errors students make in drawing graphs, as this student has done for the y-axis. Notice the x-axis is incompletely labelled since 'respiring material' is missing. Choose your scales very carefully. In the National Examination, the grid for graphs is always constructed to allow you to use more than half so that is a good way to check your chosen scale is suitable. Watch that the data may not be straightforward to plot, for example 11 and 7 here, and be careful to notice if the data does not rise in stepwise fashion. For example, 0, 2, 4, 6 are much easier to plot that 0, 1, 5, 6 which need a lot more care. Also, include a 0 point if that is appropriate, as here, and values slightly above your maximum value to be plotted (which is 11 in this question). Now the bars can be drawn making sure you leave an equal space between each, which this student has failed to do. Plot the points very carefully. See how this student has done that for seeds, earthworms and meal worms but not for woodlice. Draw all lines with a ruler and if you want to, though it is hardly necessary, you can lightly shade in the bars. If you are unhappy with your first attempt, there is always a spare grid available but a good suggestion is to work in pencil to start with. This is a very poor attempt by a weak student who has not practised the skill of drawing bar charts sufficiently so no marks would be awarded.

The bar chart below is properly drawn and you can see the difference between this, from a well practised candidate, and the work of the previous weak student. This answer would get all 3 marks.

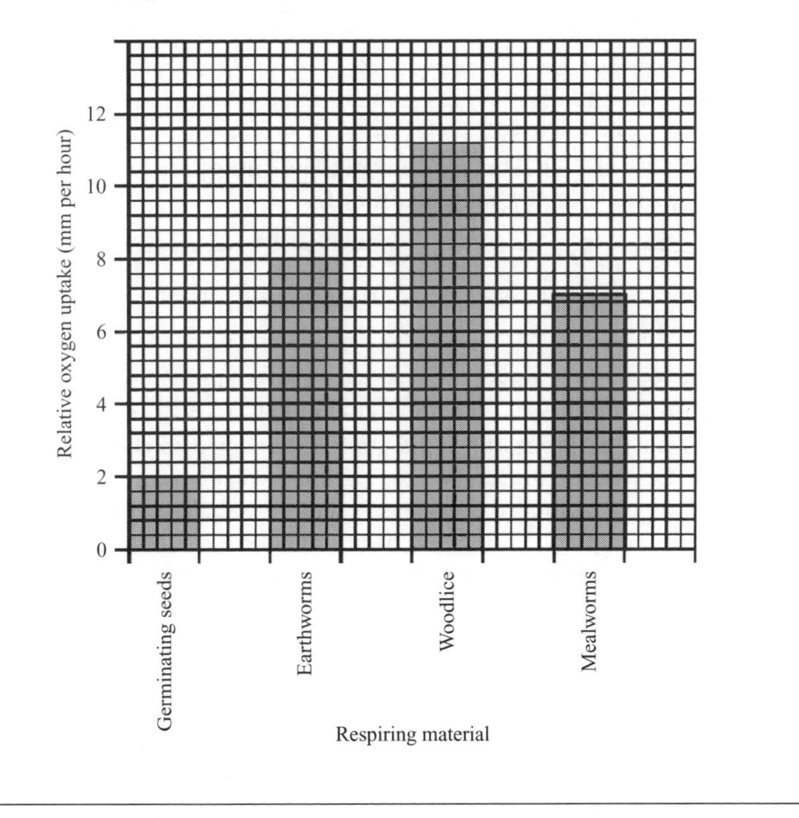

(b) Describe a control for this experiment. (1 mark)

(b) *the same apparatus but with nothing in the flask*

Biological experiments are very different in concept from those in other sciences because the material we are dealing with is living or comes from living cells. This means we have to allow for the fact that the material may be highly influenced by a number of factors, called variables, which can affect the outcome of an experiment. To prove what is causing the change, the experiment is controlled as much as possible by eliminating as many variables as possible: for example, by keeping the temperature, pH, chemical concentrations, time, etc., constant and varying only one thing at a time, such as light intensity. Here we need to think how to prove the change is caused only by the respiring material. One way of doing this is to remove it, keeping everything else the same. Even better, would be to replace the respiring material with an equal mass of something non-living, like glass beads. This candidate's answer goes part of the way but not enough. Here is a correct response, *do the experiment using the same apparatus and conditions but with no respiring tissue in the flask.* No mark awarded here.

(c) State the reason for leaving the tap open for 10 minutes before starting the measurements. (1 mark)

(c) *to give the respiring tissue a chance to settle*

Linked to the previous question, one variable which might affect this experiment is temperature. In order to let the apparatus and all the contents come to the same temperature before starting, the tap is left open so no change in volume due to the effect of the surrounding temperature will be recorded. It is generally a good idea to give biological experiments time to stabilise before starting to record data, something you should bear in mind for future questions. The candidate here has the basic idea but has not stated it precisely enough. Here is a better answer, *to allow the apparatus to come to the same temperature as the surroundings*. No mark awarded.

(d) The sodium hydroxide solution absorbs carbon dioxide. Suggest a reason for the inclusion of the filter paper in the flask. (1 mark)

(d) *increases surface area exposed to the air inside the flask*

Generally, when you see an experimental setup which includes a flat strip of paper or an equivalent, for example cutting up a tissue into small pieces, you should be thinking of how this increases the surface area. This is a concept found widely in biological systems, for example the very large surface areas to allow for gaseous exchange inside a leaf or lung. Here, using the strip of filter paper causes the exposure of the sodium hydroxide to the air to be increased. This is an excellent answer and would obtain 1 mark.

(e) (i) Draw an arrow beside the capillary tube to show the direction in which you would expect the coloured dye to move during the investigation. (1 mark)

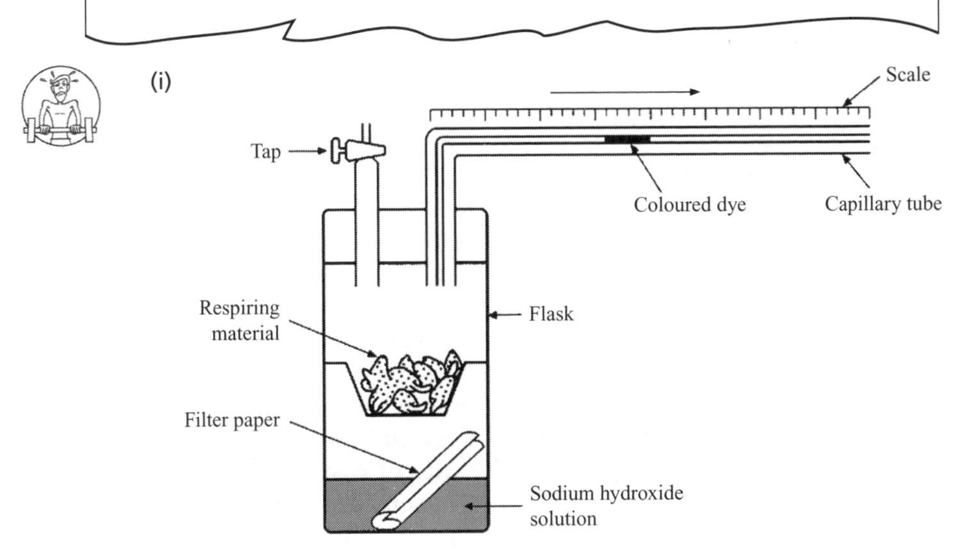

This answer is wrong and the arrow should be pointing the other way. We'll see why in the next answer. No mark awarded.

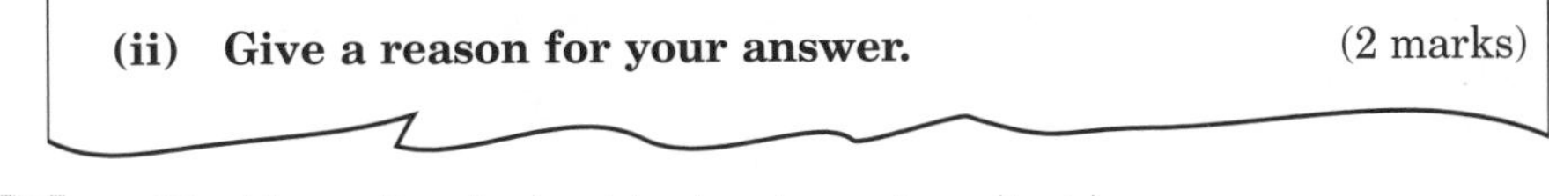

(ii) the sodium hydroxide absorbs carbon dioxide which lowers the pressure inside the flask

It can happen easily in an examination under pressure: a student can make an uncharacteristic error. The quality of the second answer here indicates a very good student so it's likely the error was, as suggested, uncharacteristic. Of course no mark can be awarded but notice how the error does not impact on the second answer which is excellent. The absorption of carbon dioxide by the sodium hydroxide (sometimes solid soda lime is used) would indeed reduce the internal pressure of the flask causing the air bubble to move towards the left. The student has given two very well worded points to gain both marks.

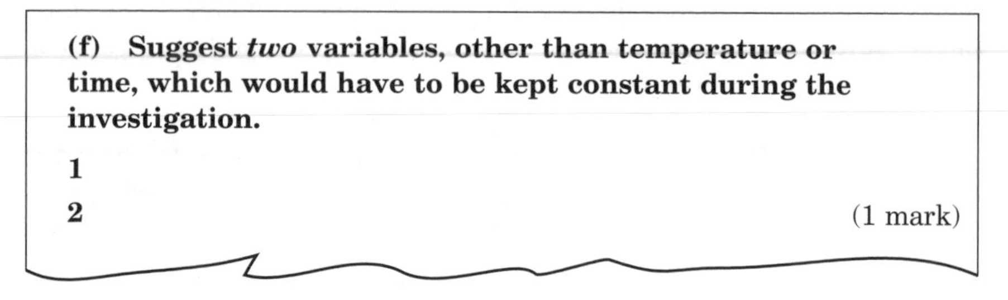

(f) Suggest *two* variables, other than temperature or time, which would have to be kept constant during the investigation.

1

2 (1 mark)

(f) *time each experiment is run*
mass of material used

This question carries only 1 mark for two answers which suggests a low level of difficulty. There are a number of variables which could be given but see how this student has missed the fact that the question actually says '.... left .... for 5 hours ...' and so *time each experiment is run* is not a satisfactory answer. The student could have stated something related to the apparatus being the same for each material or the concentration and volume of sodium hydroxide solution being the same, etc. No mark awarded.

(g) What modification would have to be made to the apparatus when measuring the rate of respiration of the green leaves of a daisy? Give a reason for your answer.

Modification (1 mark)

Reason (1 mark)

(g) *put the apparatus in a dark cupboard*
stops photosynthesis taking place

Any means of preventing light entering the system is satisfactory and the one given here is as good as any. Restricting light in experiments involving plants, especially when the question actually says 'green leaves' strongly suggests preventing photosynthesis taking place. 2 marks awarded.

4 The diagram below shows translation of part of an mRNA molecule during the synthesis of a protein.

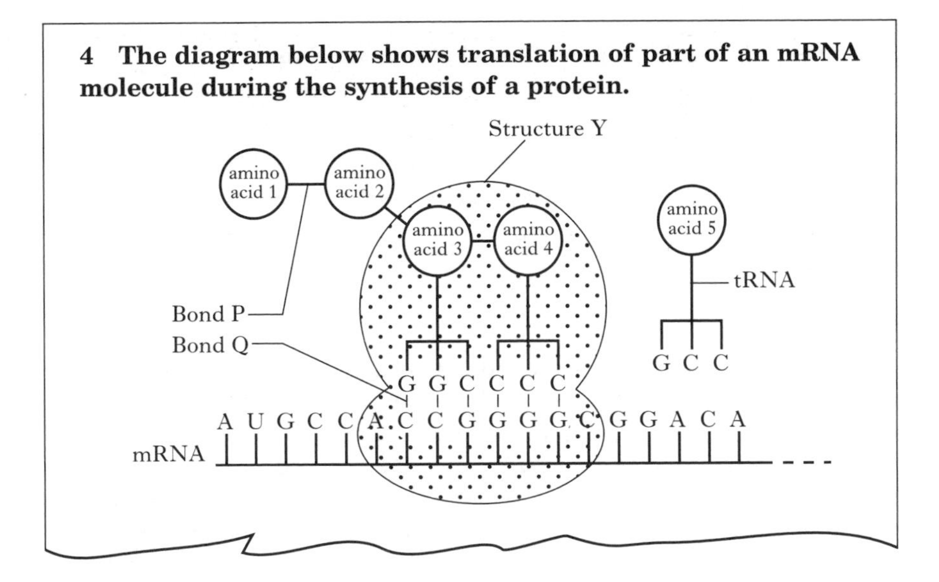

The synthesis of protein at the ribosome is another topic which lends itself well to a visual representation, such as is partially shown in the question. By using a combination of arrows and colours it will be easier to remember what is happening and in what order. Two 'types' of ribonucleic acid are mentioned here, mRNA and tRNA, and their distinctive roles need to be understood. Names such as 'transcription' (the copying of information from nuclear DNA to mRNA) and 'translation' (the reading of the mRNA by the ribosomes in the cytoplasm) are easily confused. Let's go through this question now.

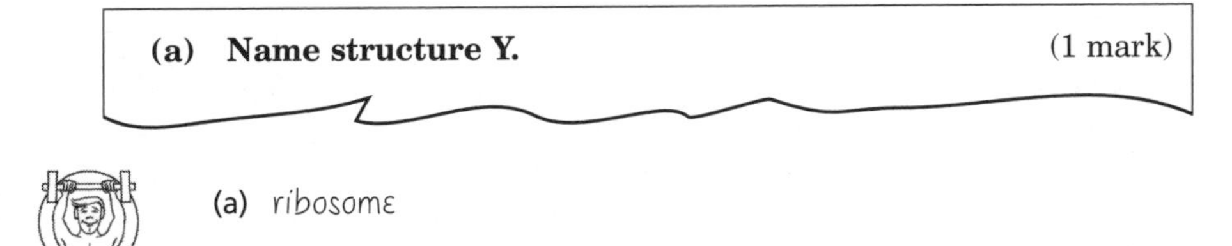

(a) Name structure Y. (1 mark)

(a) *ribosome*

A very easy knowledge-based question requiring a one-word answer. 1 mark awarded.

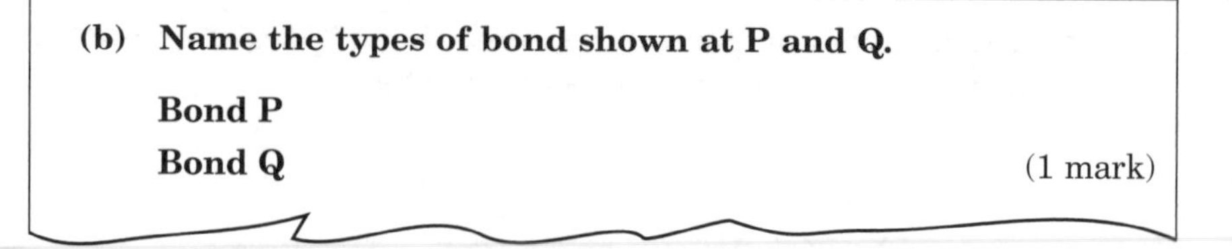

(b) Name the types of bond shown at P and Q.

Bond P

Bond Q (1 mark)

(b) peptide
covalent

This student has correctly understood the linkage between amino acids as a peptide bond. However, the student has been put off by the representation of the other linkage, probably well-known in a slightly different setting, found between the bases in the DNA molecule, a hydrogen bond. Notice, there is substantially no difference here, the linkage is between two different ribonucleic acids, tRNA and mRNA. As mentioned many times in this book, try not to be rigid in dividing your knowledge into compartments, and see connections between different parts of the course or see similarities in different representations of the same idea. No mark awarded here.

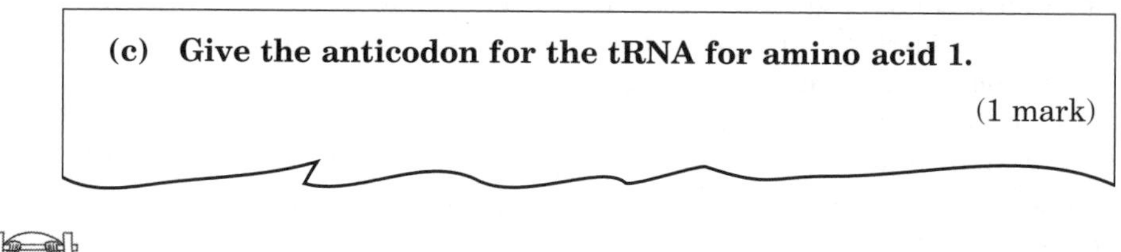

(c) Give the anticodon for the tRNA for amino acid 1.

(1 mark)

(c) UAC

This is a difficult question requiring a few steps to get to the correct answer. First, you have to understand the term 'anticodon' as the sequence of three bases found on the tRNA molecule which exactly complements a particular codon (sequence of three bases) on the mRNA. Next you have to work back along the mRNA strand to the codon which coded for amino acid 1, remembering the ribosome is moving from left to right. This codon was AUG but you are asked for the anticodon, its complement in other words, which is UAC. Remember, thymine never appears in the ribonucleic acids. 1 mark awarded here.

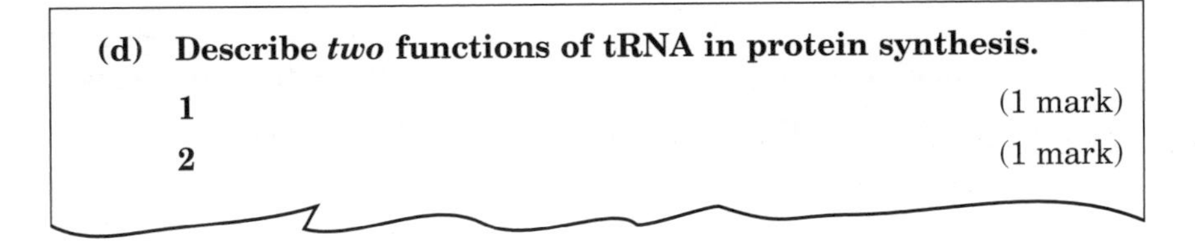

(d) Describe *two* functions of tRNA in protein synthesis.

1 (1 mark)

2 (1 mark)

(d) *picks up amino acids*
matches with bases on mRNA

The term 'transfer' suggests transporting or moving something from one place to another and indeed tRNA does transport or move amino acids. However, at Higher Grade level, you are expected to be precise in your decryption and this student, while having a superficial understanding of what's going on, has failed to be sufficiently precise. First, the tRNA does indeed pick up amino acids but in a very specific way, not randomly as this answer suggests. The answer should have read *picks up/collects/binds to specific amino acids*. Next, is the transport function of tRNA molecules carrying the specific amino acid to ribosomes for translation into protein. This answer is insufficient and would need to have been worded something like *carries/takes the amino acids to the ribosomes/site of translation/site of protein synthesis.* No marks awarded.

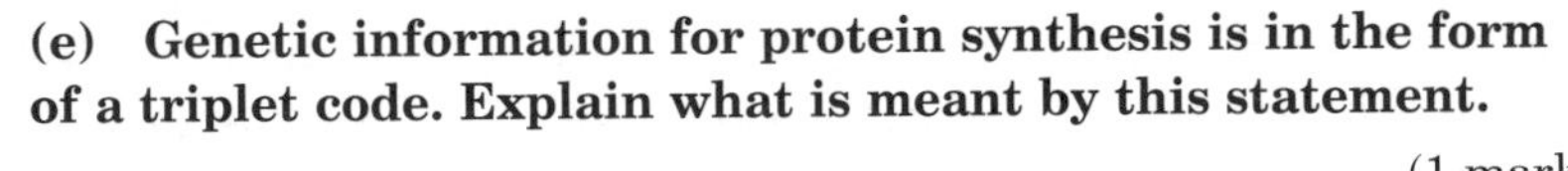

(e) Genetic information for protein synthesis is in the form of a triplet code. Explain what is meant by this statement.

(1 mark)

(e) *the code is in threes, three bases for each amino acid*

A good answer here, to obtain 1 mark.

5 A bacteriophage is a virus which attacks cells of bacteria.

The following diagram represents the life cycle of a bacteriophage.

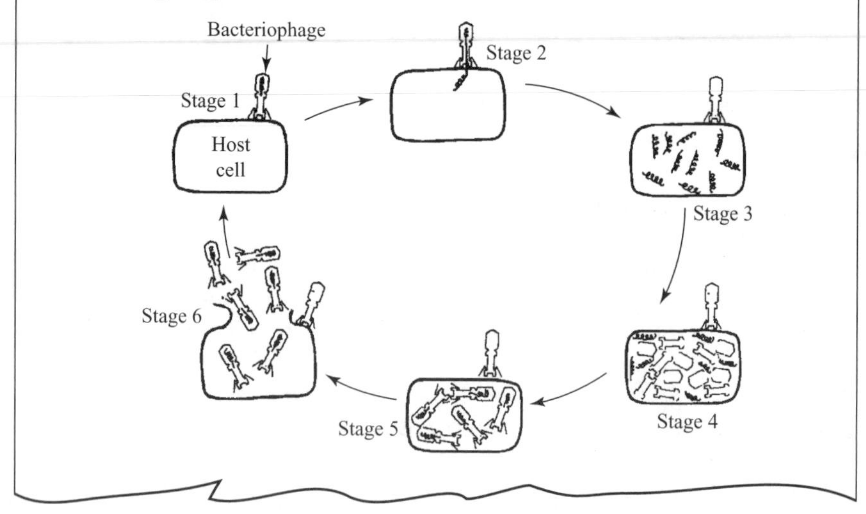

When you have any kind of cycle in Biology, diagrams usually help a lot. Think of the nitrogen cycle, Calvin cycle, Krebs cycle and how these naturally lend themselves to a visual representation to help you remember sequences and terms. That very much applies here to the life cycle of the virus which infects bacterial cells. This is called a bacteriophage. Notice the use of the ending 'phage' meaning 'eating'. This virus 'eats bacteria' and it is useful to try to remember the term in this way. An average student answered this question as follows.

(a) Name the substance which is being introduced into the host cell by the bacteriophage in stage 2. (1 mark)

(a) *nucleic acid*

This is a simple one-word answer and is correct. 1 mark awarded here.

(b) Describe what is happening during stage 3 and stage 4.

Stage 3 (1 mark)

Stage 4 (1 mark)

(b) *more nucleic acid has been injected*
viruses being produced

The use of 'describe' together with the spacings given indicates an answer of more than one or two words is needed.

In stage 3, there is the appearance of more nucleic acid but the student has a flawed understanding of what has happened. You can see no more viruses are attached to this cell at stage 3 so no more nucleic acid can be injected, which should have helped this student eliminate this description. The correct answer should have read something like *viral nucleic acid is being replicated inside the host cell*. No mark awarded.

In stage 4, viral assembly has not yet taken place so this answer is clearly wrong. In order for new viruses to be fully assembled, the protein heads need to be made first and that is what is happening here; *new viral protein heads are being made.* No mark awarded here.

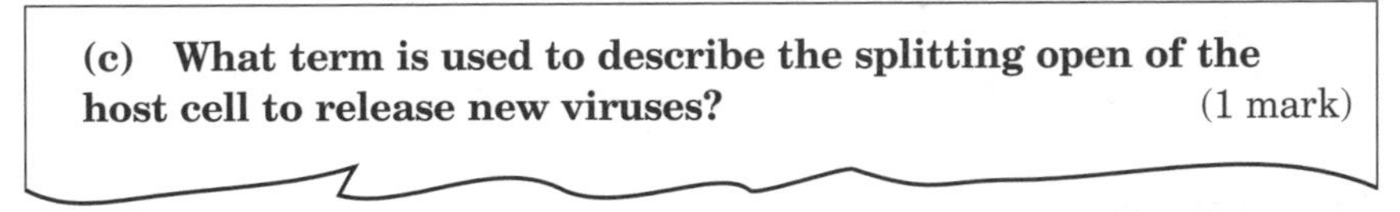

(c) What term is used to describe the splitting open of the host cell to release new viruses? (1 mark)

(c) *lysis*

Notice this question leads you to give a one-word answer by asking 'What term ....' rather than 'What terms ....'. The term 'lysis' applies in various settings such as the bursting of animal cells when water enters by osmosis. 1 mark awarded here.

EXAMPLES

Unit 2 Genetics and Adaptation

1 (a) A gene from a jellyfish can be inserted into a bacterial plasmid using a genetic engineering procedure.

Some of the stages are shown in the diagram.

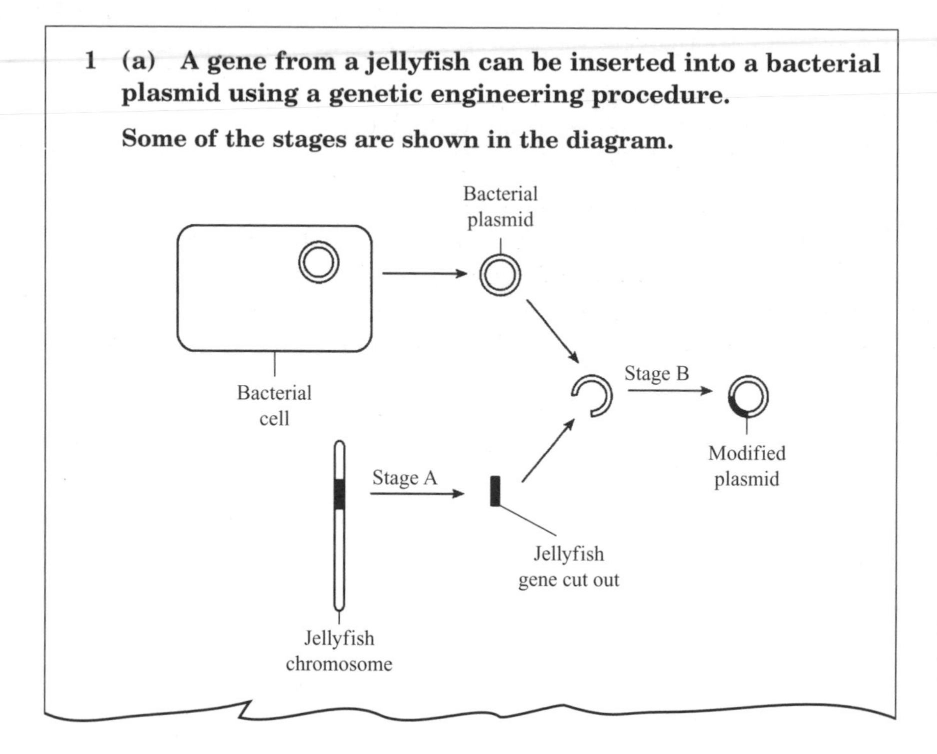

This is a question which tests your understanding of genetic terms such as plasmid and genetic engineering, as well as some of the procedures involved in modifying a bacterial cell's genetic material. Remember that bacteria have small circular structures of DNA called plasmids which can be 'cut' in places using chemical 'scissors' and have new pieces of DNA inserted. Both these processes involve enzymes and the whole procedure is often referred to as genetic engineering.

Let's see how a weak student attempted this question.

(i) Give *one* method which could be used for locating the gene in the jellyfish chromosome. (1 mark)

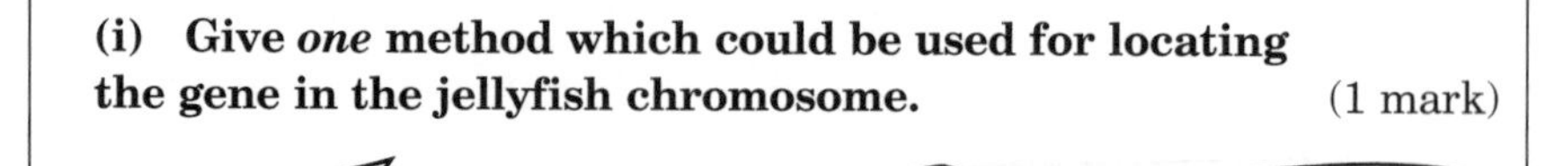

(i) *crossing over*

It would be clear to an examiner that this student really does not know the answer but may have heard of crossover values, which would have been acceptable. Crossing over is the genetic exchange of materials during meiosis and is not appropriate for this answer. Another good answer would have been *gene probe* which is a piece of DNA complementary to the gene being looked for. Such a probe can be radioactively labelled. Some alternative answers could be *chromosome mapping/gene mapping/crossover values.* No mark awarded.

(ii) Name the enzymes involved in the following stages of the genetic engineering procedure.

1 Cutting the jellyfish gene out of its chromosome (Stage A) (1 mark)

2 Sealing the jellyfish gene into the bacterial plasmid (Stage B) (1 mark)

(ii) 1 *cutting enzyme*

Notice the student has used two terms, 'cutting' and 'enzyme', both found in the question, which should have been an alert that this answer can't be correct. The correct answer is *restriction endonuclease/endonuclease*. No mark awarded.

2 *joining enzyme*

Again, note the use of the word 'enzyme'. The student knows the function of the enzyme but not the technical name, *ligase*. No mark awarded.

(b) Name *one* human hormone that is manufactured by genetically engineered bacteria. (1 mark)

(b) *interferon*

The student has correctly remembered interferon can now be made using genetic engineering but the question asks for a human hormone. Interferon is not a hormone but an antiviral protein. A correct answer could have been *insulin* or *growth hormone*. No mark awarded.

2 Duchenne muscular dystrophy (DMD) is a recessive, sex-linked condition in humans that affects muscle formation.

The diagram shows an X-chromosome from an unaffected individual and one from an individual with DMD.

X-chromosome from unaffected individual

X-chromosome from individual with DMD

Genetics is riddled with terminology which can lead to confusion. The use of 'flash cards' has been suggested elsewhere as one possible strategy for learning the terms and making sure you don't get them mixed up. Two terms very often confused are 'phenotype' and 'genotype.' Phenotype is really the expression of the genes contained in the nucleus of an organism which can often be affected by the environment. Sometimes that expression can actually be seen such as the height of a plant but sometimes not, such as the blood group of a human. The genotype can be used to mean the total genetic information contained in the nucleus or, more commonly, just the combination of alleles of a gene or genes of interest, sometimes called the genome.

Let's look at these answers from a well-prepared candidate.

(a) Using information from the diagram, name the type of chromosome mutation responsible for DMD. (1 mark)

(a) deletion

Chromosome mutations result in a change in the structure or number of chromosomes and, as such, can often be physically seen, as the one causing DMD can. Notice the affected chromosome is physically shorter than the normal one indicating a loss or deletion of genetic material. 1 mark awarded here.

(b) *On the diagram* of the chromosome from the *unaffected individual*, put a *cross* (*X*) on the likely location of the gene involved in DMD. (1 mark)

X-chromosome from individual with DMD

The candidate has noted the very clear instructions in this part to put the 'X' actually on the diagram and has hit exactly the correct bar representing the section of the chromosome which has been lost or deleted. 1 mark awarded here.

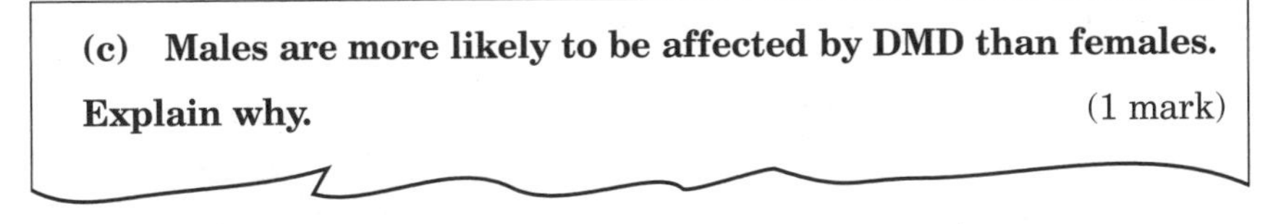

(c) Males are more likely to be affected by DMD than females.

Explain why. (1 mark)

(c) *since males are XY and females are XX, males only need one copy of the affected gene to have the condition whereas females need two copies*

Although this question carries only 1 mark, notice the spacing and also the instruction to 'explain.' These indicators tell you a one- or two-word response will not be enough. See how this candidate has carefully demonstrated a good understanding of the different genotypes of males and females and how this leads to a difference in the inheritance of sex-linked conditions such as DMD. 1 mark awarded.

(d) A person with DMD has an altered phenotype compared with an unaffected individual.

Explain how an inherited chromosome mutation such as DMD may result in an altered phenotype. (1 mark)

(d) *the normal protein cannot be made and instead an altered one is produced which may not function*

Similarly, this question needs a fuller treatment than just 'state' or 'name' and this candidate has picked up on this. Altering the chromosome by deletion will cause the protein synthesised to be changed or simply not to be made at all. Although this change is at the biochemical level as such and cannot be 'seen', it is still a change in phenotype. 1 mark awarded.

3 Today, a part of East Africa contains a variety of *Cichlid* fish species. The drawings show the evolution of a group of these *Cichlid* species together with information on their food sources.

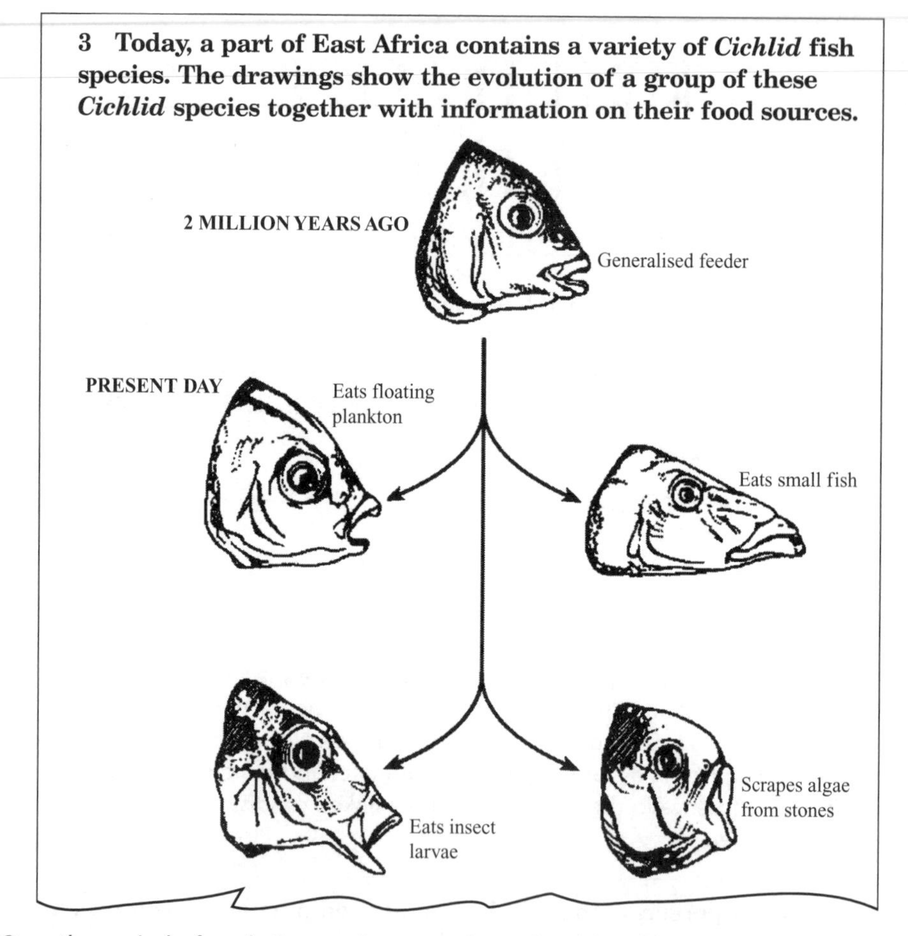

Over the period of evolutionary time, a variety of species can arise from one ancestral species. The new variety can move into various ecological niches to exploit these for food and so forth. This is called adaptive radiation. Here are a good candidate's answers to this question.

(a) Explain how the information above illustrates adaptive radiation. (1 mark)

(a) four new species have arisen from one common ancestor, each able to exploit different niches

Notice how this candidate has given a full answer here, taking note of the need to 'explain' and also the spacing given to get the mark.

(b) What advantage does each *Cichlid* species gain from being a specialised feeder? (1 mark)

(b) since each species is adapted for different feeding they won't compete for the same food

Again, the candidate has given a full answer, recognising the consequences of each new species having different mouth parts, thereby enabling them to eat different foods and therefore not having to compete for the same resources. 1 mark awarded.

4 (a) The diagram represents a plant with two regions magnified to show tissues involved in transport.

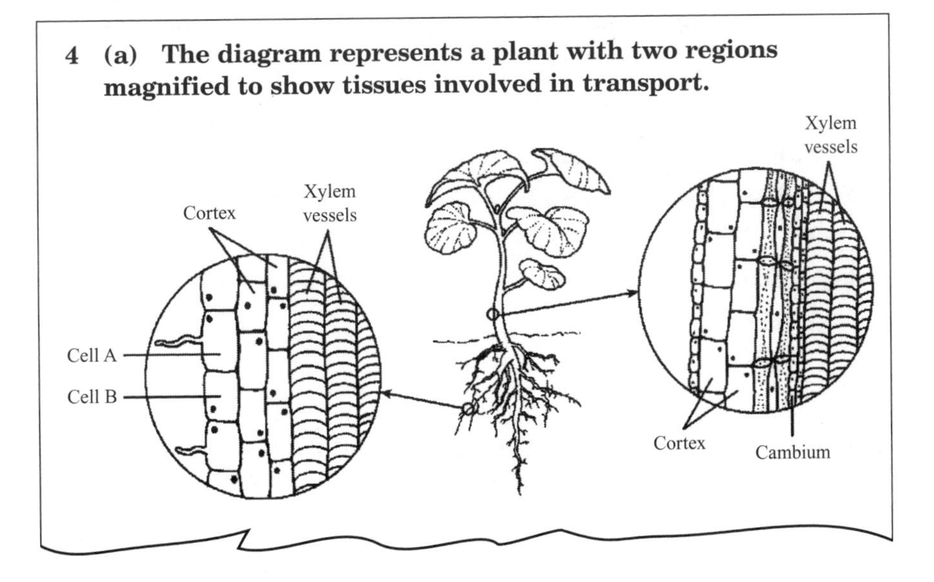

The relationship between structure and function applies widely in Biology at the level of individual cells as well as whole organisms. Think of the shape of a red blood cell and how it is adapted for its function of oxygen transport. You would do well to list some examples from both plants and animals to illustrate this important relationship which could easily form the basis of an extended type response question. Here is an average student's answers for this question.

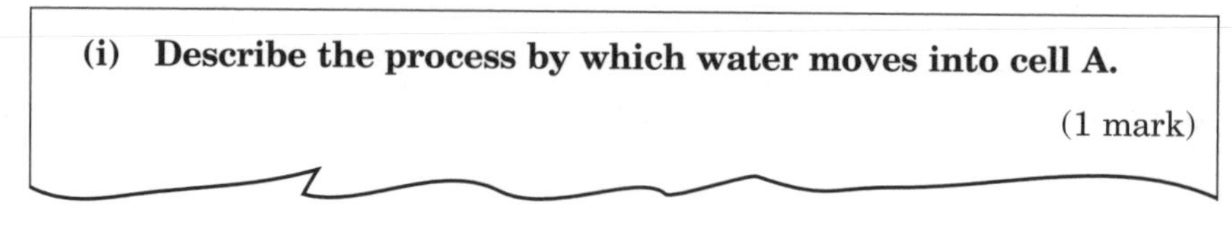

(i) Describe the process by which water moves into cell A.

(1 mark)

(i) osmosis

Notice this question asks for a description and the spacing suggests a more extensive answer than this student has given. You have to bring out the difference in the water concentration, the so-called concentration gradient, which exists between the outside and inside of cell A. This student has not described the process, merely named it. No mention of the gradient or that the water molecules are in highest concentration outside is given. You need to keep in mind that the expectation at Higher Grade is much greater than Standard Grade and answer accordingly. A correct answer might read *water is found in high concentration outside cell A compared to inside the cell and so moves down the concentration gradient into the cell.* No mark awarded.

(ii) Cells A and B have a similar function.

Explain how the structure of cell A makes it better adapted to its function than B. (1 mark)

(ii) cell A has a large surface area

You can see from the diagram that the root hair cell is well-adapted to its function but the function is not mentioned in this student's answer. The link between these two features, structure and function, comes up regularly in the National Examination so make sure you fully understand the concept. Think of the structure of a knife and a fork and how each structure is related to its function. This student has partly answered the question but not enough to gain both marks. A full answer might read *cell A has a large surface area which allows it to absorb water efficiently.* No mark awarded.

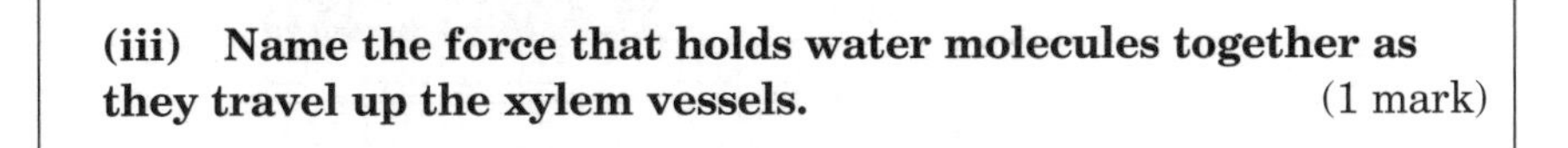

(iii) Name the force that holds water molecules together as they travel up the xylem vessels. (1 mark)

(iii) *cohesion*

Cohesion is the tendency of the water molecules to stick together so that the long thin water column is not broken. 1 mark awarded.

(iv) Cell division in the cambium produces new cells which then elongate and develop vacuoles.

Describe *two* further changes that take place in these cells as they differentiate into xylem vessels.

1 (1 mark)

2 (1 mark)

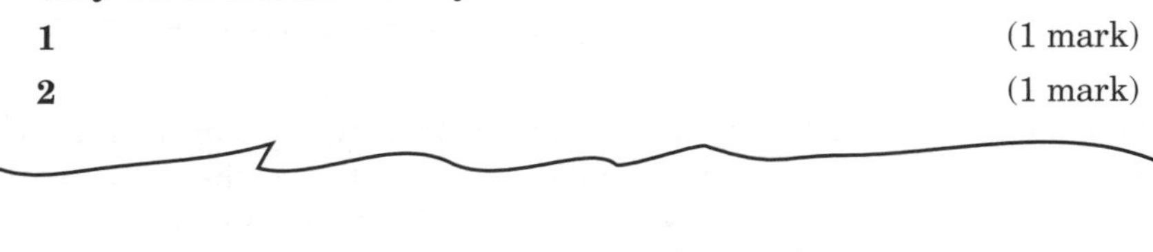

(iv) 1 *lignin is added* 2 *end walls disappear*

There are several changes which occur as the cambial cells change into xylem. These include the disintegration of the nucleus, loss of the cytoplasm and cell contents. In addition, the two mentioned by this student are correct so both marks awarded.

5 (a) The graph below shows changes in stomatal width over a 24-hour period of a plant species that is adapted to live in a hot climate.

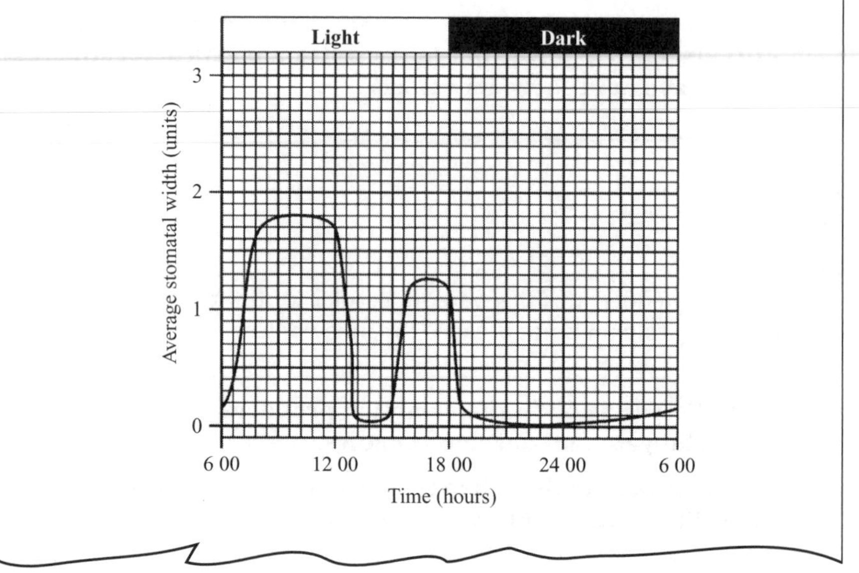

This is a slightly difficult question testing stomatal function. Stomata are the openings for water vapour, oxygen and carbon dioxide to exchange between the plant and the atmosphere. When a plant is photosynthesising, the stomata are normally open to allow for the uptake of carbon dioxide but this also leads to an increased loss of water by evaporation. This can create a problem of balancing water loss from the leaves with the uptake by the roots. If there is an imbalance such that the water loss exceeds water uptake, the guard cells surrounding the stomata start to lose turgor, a term you need to know and understand. Turgor refers to the water content of a plant cell. The more water present, the greater the turgor. Here is a weak candidate's attempt at this question.

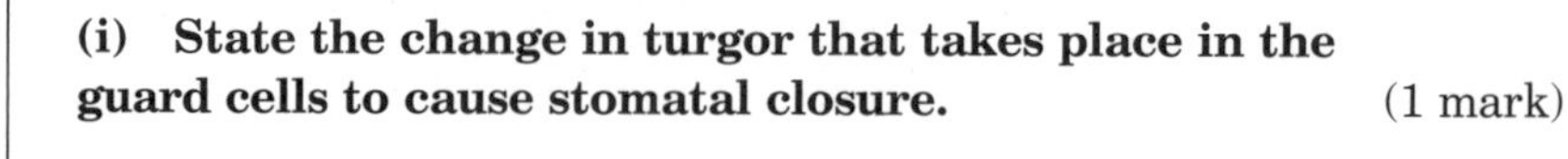
(i) State the change in turgor that takes place in the guard cells to cause stomatal closure. (1 mark)

(i) cells lose water

Notice this question asks for a statement of the change, not an explanation which is what this candidate has given. The correct answer is *decrease* or *becomes flaccid*, which is a state of lacking firmness due to a low water content. No mark awarded here.

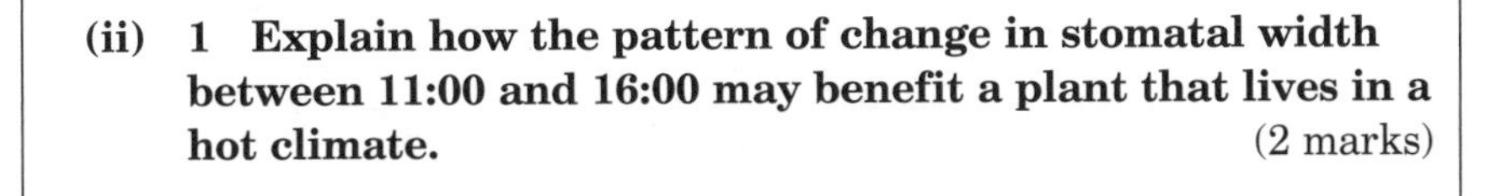

(ii) 1 Explain how the pattern of change in stomatal width between 11:00 and 16:00 may benefit a plant that lives in a hot climate. (2 marks)

(ii) 1 stomata close when it gets hot to stop plant losing water

This question is weighted with 2 marks which suggests two points are needed. Also, the spacing confirms that a fuller answer than the one given is required. Remember that this plant lives in very hot conditions and during late morning/early afternoon the temperature will be particularly high so, as you can see from the graph, the width of the stomata falls rapidly to help prevent water loss by evaporation. By about 13:30, the stomata are almost closed but open rapidly again later in the afternoon. This response ensures the plant does not lose too much water. The weak candidate has given only a partial explanation but would probably earn 1 mark. A better response might have been *the stomata close during the hottest part of the day to prevent water loss by evaporation*.

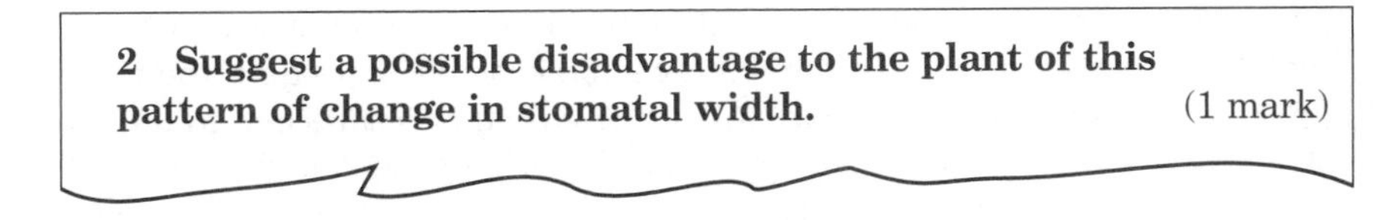

2 Suggest a possible disadvantage to the plant of this pattern of change in stomatal width. (1 mark)

2 plant can't get enough carbon dioxide

When the stomata are closed, the uptake of carbon dioxide needed for photosynthesis is greatly reduced. This answer would get the mark.

6 African wild dogs are social animals that hunt in packs. They rely on stamina to catch grazing prey such as wildebeest.

The table shows the effect of wildebeest age on the average duration of successful chases and the percentage hunting success.

Wildebeest age	Stage	Average duration of successful chases (s)	Hunting success (%)
up to 1 year	calves	20	75
from 1–2 years	juveniles	120	50
over 2 years	adults	180	45

Animals which feed in groups, such as wolves, obtain many benefits over those which are solitary feeders, such as some sharks. Feeding in a group increases the chance of a kill and prevents other animals from stealing the dead prey, helps conserve energy, feeds weaker or younger members of the group, helps prevent predators themselves being damaged, allows large prey to be brought down, and so forth. Energy conservation surfaces widely in Biology and you should be able to see this in action elsewhere, other than social feeding. For example, the number of offspring an animal might rear will be linked to the energy required by the parents to raise them. This question touches on various aspects of social feeding using the wild dog and their prey as an example. Let's go through the answers, improving on those which are either wrong or insufficient to obtain full marks.

(a) Describe the effect of wildebeest age on the average duration of successful chases. (1 mark)

(a) *time taken gets longer*

It is very common to be asked to consider the effect of changing one variable, such as temperature, on another, such as enzyme activity. When you are asked this type of question, make sure you include the two variables concerned and give a 'direction', in other words, what happens when one variable increases (or decreases). This student has failed to make the link between the wildebeest age and the time for chases. A correct response would be, *as the wildebeest age increases, the average length of time for successful chases increases*. It could also be stated this way, *as the wildebeest age decreases, the average length of time for successful chases decreases*. Notice how this answer includes a full expression of the variable, 'average duration of the successful chases' from the table. This is good practice. No mark awarded here.

(b) How many times longer does it take the wild dogs on average to successfully hunt adult wildebeest rather than calves?

Space for working **times**

(1 mark)

(b) 9

A straightforward calculation using the data from the table. 180 s for chasing adults compared with 20 s for chasing calves is an increase of 9 times.

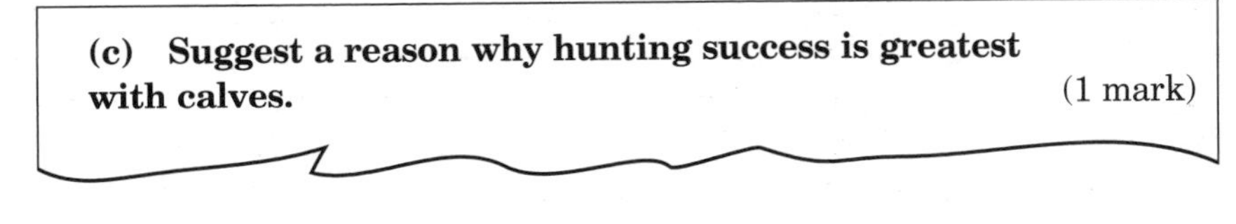

(c) Suggest a reason why hunting success is greatest with calves. (1 mark)

(c) easy to catch

Calves, by the nature of being young, will have poor stamina so can't run as fast for as long as adults. Also, their muscular and nervous co-ordination will not be fully developed, making it difficult for them to co-ordinate their escape. They may also not have learned to recognise dangers as adults do. Any of these would be acceptable answers stated perhaps as *younger animals are not able to run away as quickly as adults* or similar words. This student has made a correct statement but it is essentially just repeating the question, a common error on the part of some students. No mark awarded.

(d) Wild dogs kill a greater number of adult wildebeest than calves.

Explain this observation in terms of the economics of foraging behaviour. (1 mark)

(d) *less energy is lost*

Foraging behaviour, the searching for food, is closely linked to conserving energy. A cheetah can only run fast for so long and if it doesn't bring down its prey, the energy expended is wasted. If animals hunting in packs can bring down a large prey, such as an adult wildebeest, the food supply obtained is greater, therefore the energy expended, which is distributed over a large number of animals and not one predator, is more than compensated by the energy obtained from the kill. This student has some idea of this but has not explained it well enough. A better answer might be *the energy obtained by killing an adult animal is much greater than that obtained from killing a calf*. No mark awarded.

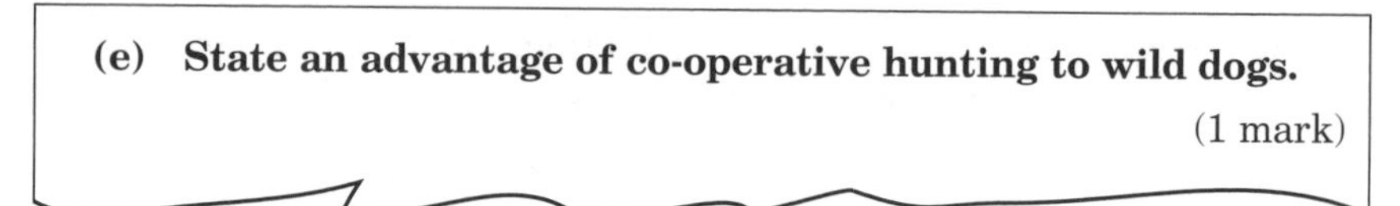

(e) State an advantage of co-operative hunting to wild dogs. (1 mark)

(e) *animals hunting together can kill bigger prey*

Linked to the preamble above, you can see the reasons and advantages for animals to hunt in groups or packs. This answer is one of many, such as *energy output per individual is less* or *can tire out prey* and is perfectly acceptable. 1 mark awarded.

(f) Following a successful hunt, wild dogs may be displaced from their kill by spotted hyenas. What type of competition does this show? (1 mark)

(f) *interspecific*

The distinction between words which begin with the prefix 'intra' or 'inter' can be confusing. An easy way perhaps is to think of a train journey between two different cities. What kind of train would you get? An intercity surely! Try to keep this generalisation in mind when dealing with such words as 'intraspecific' and 'interspecific', meaning 'between members of the same species' and 'between members of different species' respectively. This question requires only a one-word answer which is correct here. 1 mark awarded.

EXAMPLES

Unit 3 Control and Regulation

1 The diagram shows parts of the chromosome in the bacterium *E. coli*. The list has three molecules involved in the genetic control of the lactose metabolism.

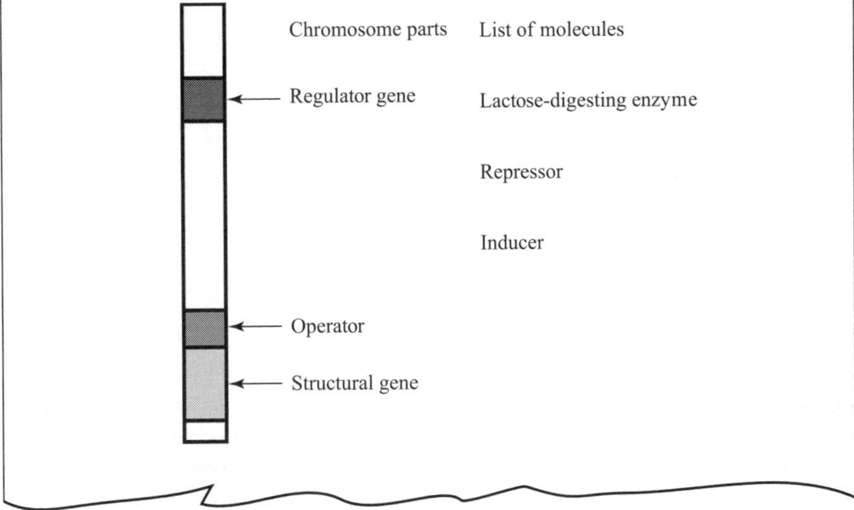

The model proposed in the 1960s by Jacob and Monod causes problems for students. This is a section of the work which lends itself well to a visual representation of the events involving genetic control, in other words a large and well-labelled diagram.

Here is an average pupil's response to this question.

(a) Complete the table by writing *TRUE* or *FALSE* in each of the spaces provided.

Statement	True /False
The repressor can bind to the operator	
The structural gene codes for the repressor	
The inducer can bind to the repressor	
The regulator gene codes for the lactose-digesting enzyme	

(2 marks)

(a)

Statement	True /False
The repressor can bind to the operator	*T*
The structural gene codes for the repressor	*T*
The inducer can bind to the repressor	*F*
The regulator gene codes for the lactose-digesting enzyme	*F*

Notice in the question stem it tells you clearly to use the words **TRUE** or **FALSE** yet this pupil has used letters instead. This would not be penalised but it is careless. Try to discipline yourself always to answer in the style asked for so that, where it might matter, you won't go wrong and lose marks. Also, 4 answers are required for 2 marks so a sliding scale would be used here of 4 correct answers for 2 marks, 2 or 3 correct answers for 1 mark and anything less gets nothing.

True is correct here. The repressor molecule does indeed bind to the operator gene.

True is not correct here. The structural gene codes for the enzyme not the repressor.

False is not correct here. The inducer can bind to the repressor.

False is correct here. The regulator gene codes for the repressor not the enzyme molecule.

This pupil has got 2 out 4 correct and would be awarded 1 mark.

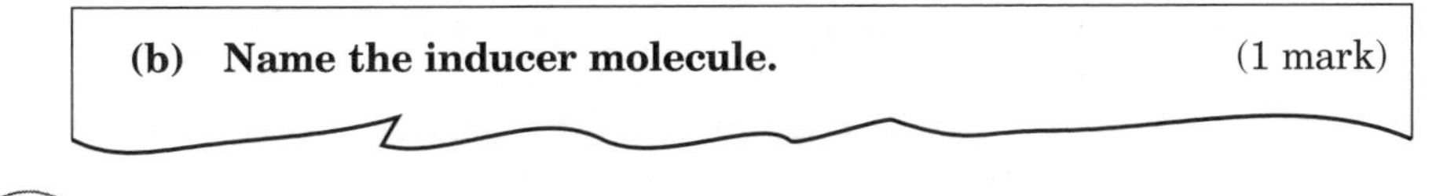

(b) Name the inducer molecule. (1 mark)

(b) sugar

The inducer for this control system is actually lactose. Notice the question has a slight clue to this by giving the name in 'lactose-digesting' so providing you know that the enzyme is acting on the inducer, you would be able to work out it must be lactose. Sugar is insufficiently precise so no mark would be awarded.

(c) Give *one* advantage to *E. coli* of having this type of genetic control system. (1 mark)

(c) no enzyme is made when it's not needed

This answer is one of many and is absolutely fine. Other answers might be *saves energy, conserves resources, does not make enzyme when lactose is absent.* Notice it can be stated conversely, i.e. *enzyme is made when it's needed.* Make sure you give only one answer as clearly asked for in the question. 1 mark would be awarded here.

2 An investigation was carried out to compare photosynthesis in oak and nettle leaves.

Six discs were cut from each type of leaf and placed in syringes containing a solution that provided carbon dioxide.

A procedure was used to remove air from the leaf discs to make them sink. The apparatus was placed in a darkened room. The discs were then illuminated by a lamp covered with a green filter. Leaf discs which carried out photosynthesis floated.

The positions of the discs 1 hour later are shown in the diagram below.

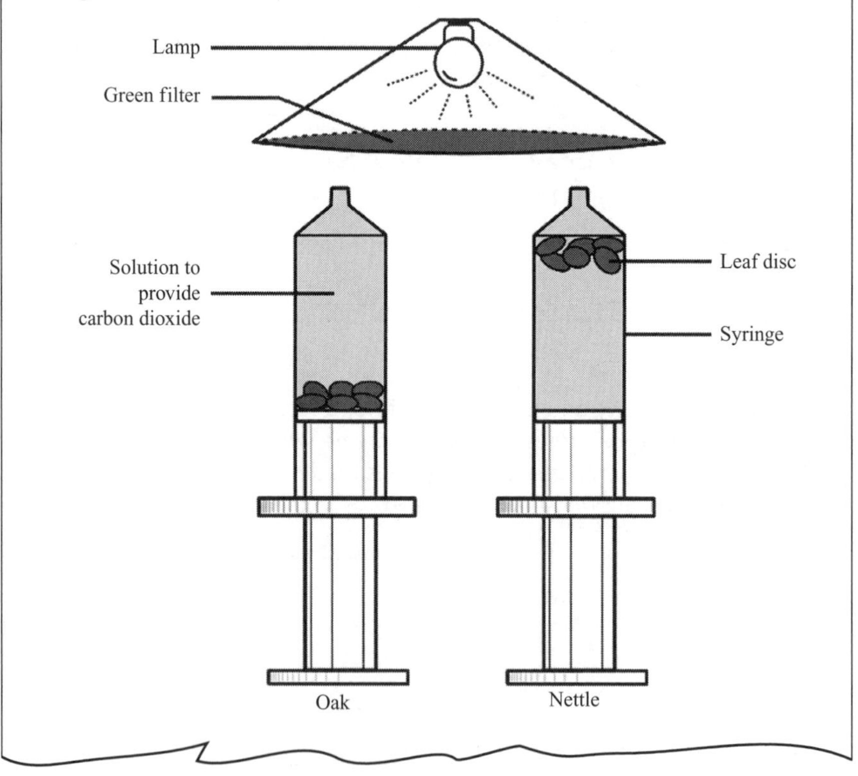

Another of the questions on an experimental procedure which you may not have carried out yourself. The graphical display of data you are asked to draw is in the form of a line graph along with linked questions to photosynthesis and sun and shade plants.

(a) Suggest a reason why the investigation was carried out in a darkened room. (1 mark)

(a) so that the only source of light for the discs is the lamp

Here we are comparing two different plants so we try to keep everything else constant. One variable is light and since we want the discs to be uniformly illuminated with green light, we need to keep out other sources, hence the dark room. This candidate has answered this question very well. 1 mark awarded.

(b) Explain why it was good experimental procedure to use six discs from each plant. (1 mark)

(b) to get a more accurate result

A common error, mentioned elsewhere, is the confusion between accuracy and reliability. Here, by using six discs from each plant instead of, for example, just one, we don't increase the accuracy but, we do increase the reliability of the results. Generally, by increasing the sample size, results become more reliable. A correct answer here would be *to make the results more reliable* or *to reduce effects of unusual results.* No mark awarded.

(c) In setting up the investigation, precautions were taken to ensure that the results obtained would be valid.

Give *one* precaution relating to the preparation of the leaf discs and *one* precaution relating to the solution that provided carbon dioxide.

Leaf discs (1 mark)

Solution that provided carbon dioxide (1 mark)

(c) the same cork borer would be used to cut the discs so that they would all be the same size

the amount of solution used for each would be the same

There are several responses to each of these questions but you are asked for only one for each. Using a cork borer would indeed ensure uniformity of the leaf discs and increase validity. 1 mark awarded.

At Higher Grade level, *amount* is not a satisfactory term because it is imprecise. You need always to use the correct quantitative terms such as 'volume', 'concentration', 'mass', etc. Had this student said *the volumes of the solutions used for each would be the same*, a mark would have been awarded.

(d) Suggest a reason why the leaf discs which carried out photosynthesis floated. (1 mark)

(d) *bubbles of gas given out make the discs float*

As photosynthesis takes place, oxygen gas is produced and as it is attached to the leaf discs, it will cause them to become less dense and so will float up the surface. Although this student has not stated the gas is oxygen, the answer is still satisfactory. 1 mark awarded.

(e) Nettles are shade plants which grow beneath sun plants such as oak trees.

Explain how the results show that nettles are well adapted as shade plants. (2 marks)

(e) *the discs still float even under green light*

This question touches on the differences between plants which grow in sunny or shady conditions. Within a forest, for example, are layers of plants with the tree canopies at the top getting the most light. Plants beneath the trees need to be able to grow in shady conditions and often can use light which the trees above don't use. Since light is a spectrum of colours, green light will be what is transmitted to the nettles below the oaks, which have already absorbed most of the other colours. Nettles grow very well in shady conditions because they can utilise low light intensities but also green light which this experiment demonstrates well. It is quite common for plants to be able to do this, using wavelengths of light different from those used by other plants to prevent interspecific competitition. This student clearly has not understood this point and the answer also demonstrates a lack of awareness of the weighting and spacing for an explanation. A correct answer might read *after the oak trees have absorbed most of the colours of white light, green light will be transmitted to the nettles below. The nettle discs can still photosynthesise even under green light.* No marks awarded.

(f) What name is given to the light intensity at which the carbon dioxide uptake for photosynthesis is equal to the carbon dioxide output from respiration? (1 mark)

(f) compensation point

A straightforward question requiring a very short answer which is correct here. 1 mark awarded.

(g) In another investigation, the rate of photosynthesis by nettle leaf discs was measured at different light intensities. The results are shown in the table.

Light intensity (kilolux)	Rate of photosynthesis by nettle leaf discs (units)
10	2
20	26
30	58
40	89
50	92
60	92

Plot a line graph to show the rate of photosynthesis by nettle leaf discs at different light intensities. Use appropriate scales to fill most of the graph paper.

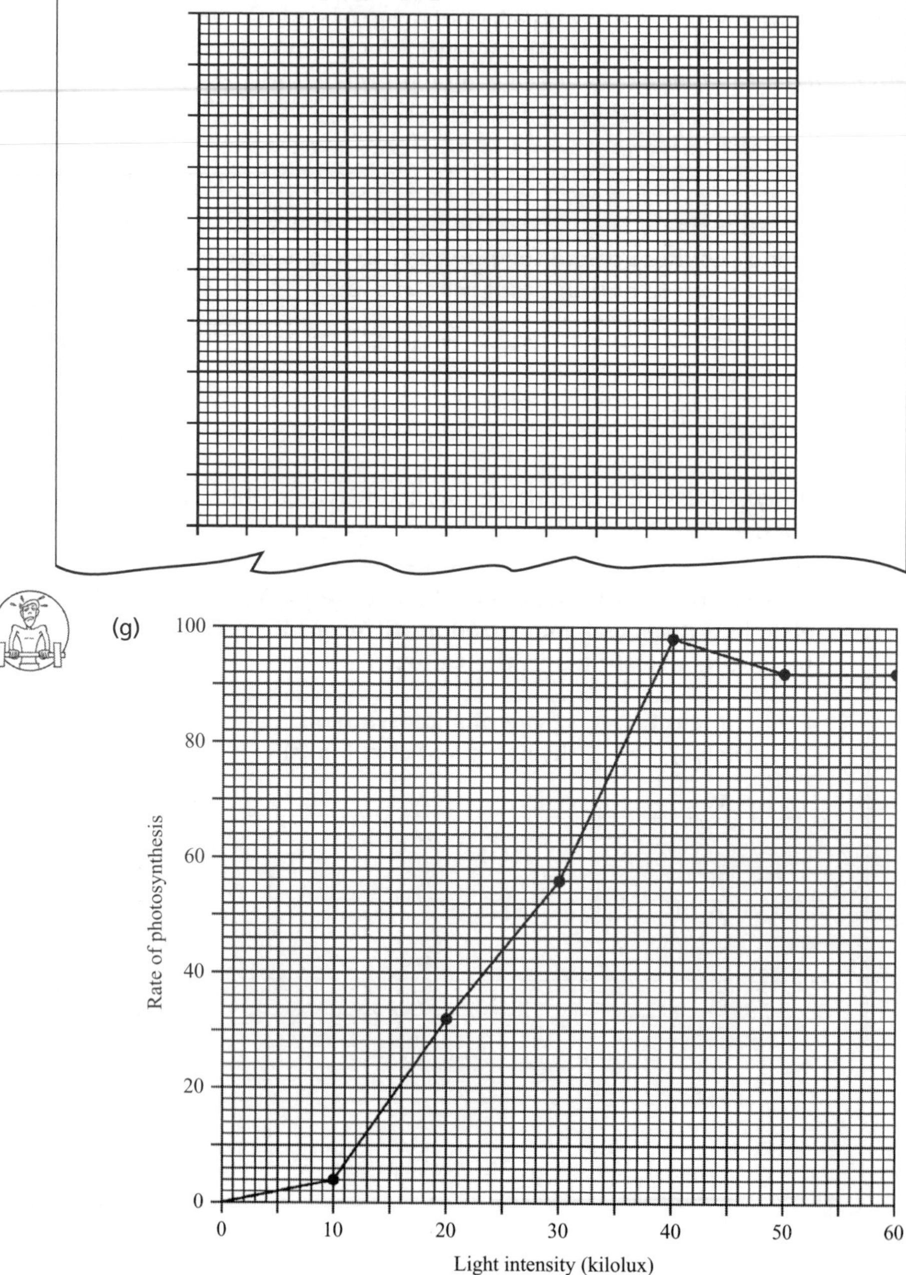

Drawing good graphs is very often a problem for students and something which needs practice to make sure you get everything which is needed down on the paper. Here is a reminder of the features of a good line graph. Make sure each axis is properly labelled and the units, where they are given, are included. Choose suitable scales for each axis, and ensure that half or more of the grid is used. Plot the points very carefully; it is easy to make a mistake here so take your time. Now join the plotted points neatly with a ruler, not free-hand. Do not plot points for which you have no data, for example, zero values. Make sure you plot the independent variable on the x-axis and the dependent variable on the y-axis. This student's attempt is very poor. Notice how the scale is correct for each axis and the variables appropriately placed on the x and y axes. However, the labelling of the y-axis is incomplete and should read *rate of photosynthesis by nettle leaf discs [units]*. Next, the actual plotting is almost completely wrong. The student has not taken enough time to work out that each of the smallest boxes represents 2 units on the y-axis. Also, the graph has been drawn backwards to the origin (0,0) but no data was given for this. No marks awarded here but let's see what would get full marks and why. Look at this plot by a good student.

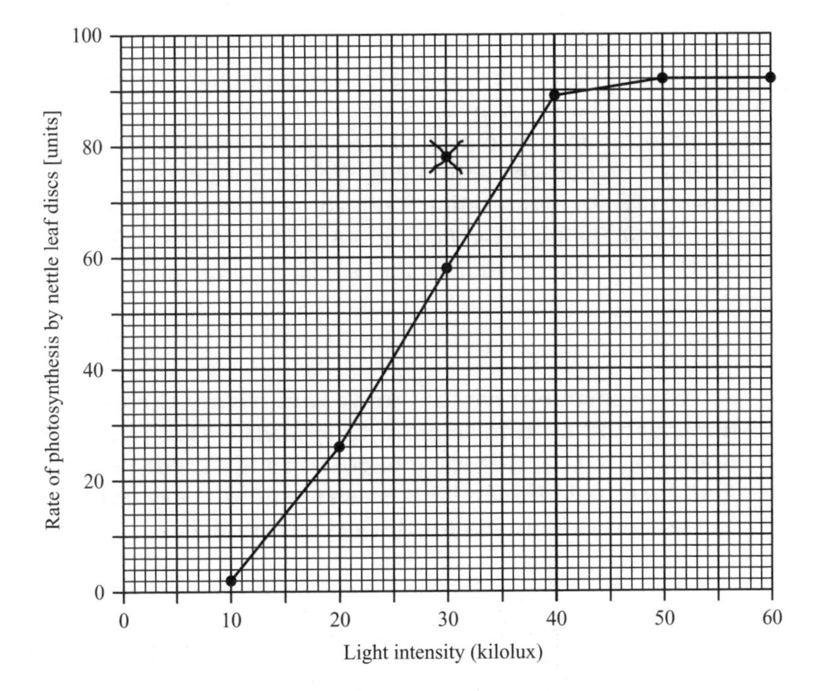

Notice how all the points are correctly plotted except for one at 30 kilolux of light intensity. However, the student has made it very clear to the examiner that this plot was not intended by crossing it out. Everything else is absolutely spot on and would get both of the possible marks.

> **(h) From the table, predict how the rate of photosynthesis at a light intensity of 50 kilolux could be affected by an increase in carbon dioxide concentration.**
>
> **Justify your answer.**
>
> **Effect on the rate of photosynthesis**
>
> **Justification** (1 mark)

(h) *it would increase*

more carbon dioxide increases photosynthesis

This last part touches on the concept of limiting factors which has been mentioned elsewhere. At 50 kilolux of light intensity and above, there is no further increase in the rate of photosynthesis so light intensity is not the limiting factor here. An increase in another variable, such as carbon dioxide concentration, would most likely increase the rate of photosynthesis, though there is no data to confirm this. While this student has made a correct prediction, the justification is not correct and no marks would be given. Had the student written *light intensity is no longer a limiting factor so carbon dioxide concentration could be a limiting factor* the mark would be given.

> **3 An experiment was set up to investigate the effect of photoperiod on flowering in *Chrysanthemum* plants. Four plants A, B, C and D were exposed to different periods of light and dark in 24 hours. This was repeated every day for several weeks and the effects on flowering noted.**

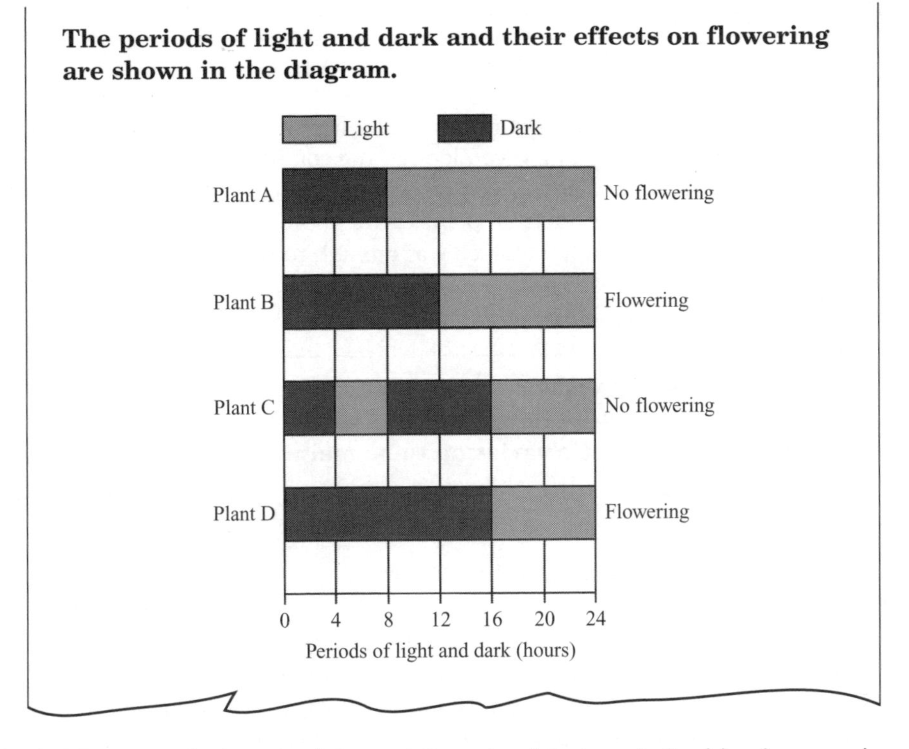

The periods of light and dark and their effects on flowering are shown in the diagram.

The link between the length of day and flowering (photoperiodism) has been made elsewhere but this question needs you to read data presented in a slightly different way. The diagram plots the effect of different periods of light on flowering in *Chrysanthemum* plants. You need to take a few minutes to work out what is being shown here. The question then leads on to the link between photoperiodism and mammalian behaviour. Light plays a huge part in the biological world, influencing plants and animals in many different ways. It is important that you appreciate the diversity of the situations and be able to draw on different examples appropriately. Let's see what a weak pupil made of this question.

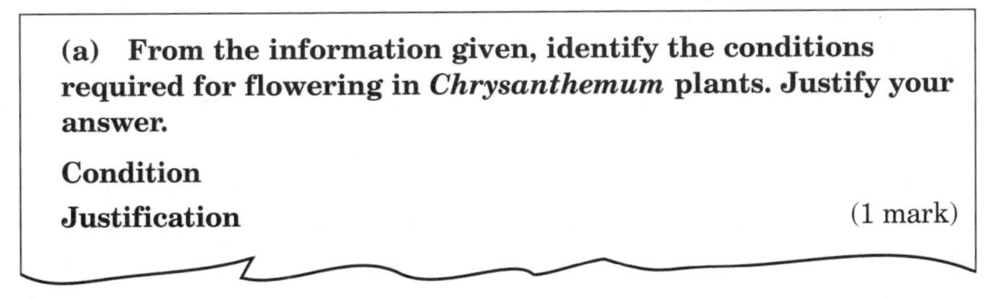

(a) From the information given, identify the conditions required for flowering in *Chrysanthemum* plants. Justify your answer.

Condition

Justification (1 mark)

(a) *no light for 12 hours*

no flowering happens if less

Notice your answer has to use the information given which this student has done but not precisely enough. If you look at the conditions when flowering does take place in plants B and D, you can see in each case continuous darkness was present for at least 12 hours. In plant C, the photoperiod of 12 hours has been broken by an hour of light which was enough to prevent flowering. A correct response might read *continuous/constant darkness for at least 12 hours because if this is interrupted flowering will not occur*. No mark awarded.

(b) Flowering response to photoperiod ensures plants within a population flower at the same time. Explain how this enables genetic variation to be maintained. (1 mark)

(b) *more plants would be pollinated*

Again, this student has some insight into the consequences of all the plants flowering at the same time but has not gone far enough. Remember that flowers are the sexual organs of plants and necessary for reproduction to take place. For any one type of plant, the more flowers that are open and mature at the same time, the better the chances of mixing up genetic material from very different varieties of the plant and producing new variations. A better answer would be *gives much more chance of pollen from one variety to fertilise an ovule from another plant.* No mark awarded here.

(c) Mammals also show photoperiodism.

Describe how one type of mammal behaviour can be affected by photoperiod. (1 mark)

(c) *birds migrate in the autumn and winter*

This answer is correct but not for this question which asks about mammals not birds. Notice the importance of reading questions carefully so that you answer what is being asked, not what you think is being asked. There are many options here such as *the snowshoe hare's fur turns white as the day length decreases*. No mark awarded.

4 The graph shows the results of an investigation into the relationship between environmental temperature and body temperature for a bobcat and a rattlesnake.

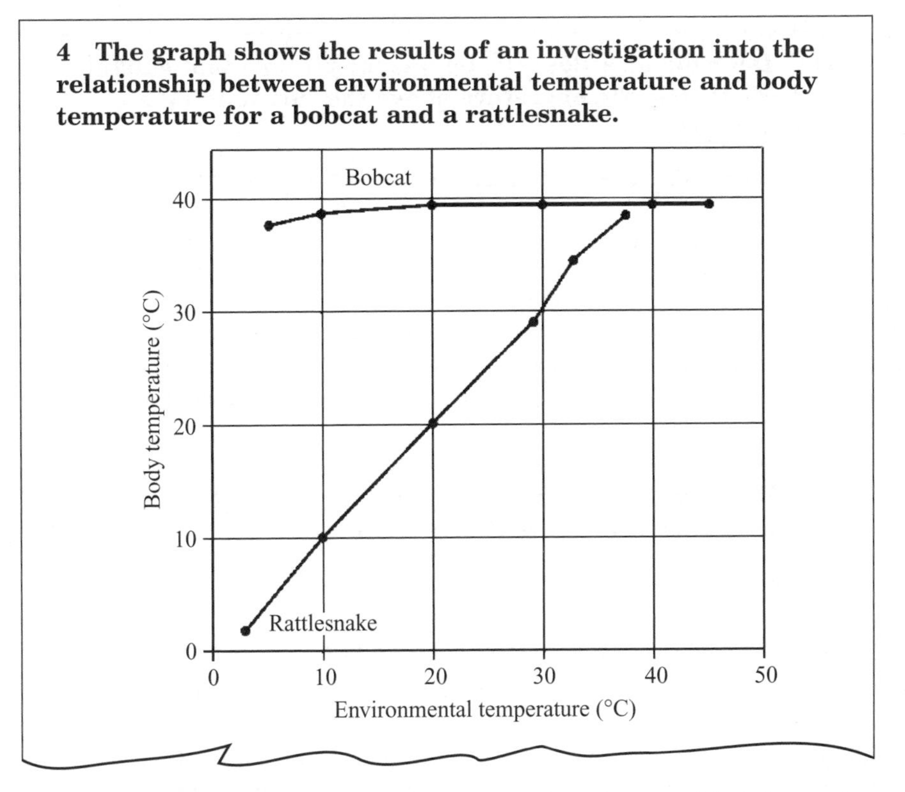

This question is a variation on the previous one by using graphical data rather than data in the form of a table. It is good to be able to manage different ways of presenting similar results. Notice you don't really need to know much about the animals themselves so don't be put off if you come across a named animal you have never heard of. There is enough information here for you to get full marks as this excellent candidate did.

(a) Using information from the graph, underline one of the alternatives in each pair to make the sentence correct.

The rattlesnake is an [ectotherm/endotherm] because the results show that it [can/cannot] control its body temperature.

(1 mark)

(a) *The rattlesnake is an [ectotherm/endotherm] because the results show that it [can/cannot] control its body temperature.*

This is a correct answer but notice both are needed for the award of the mark as here.

(b) Describe a rattlesnake behaviour pattern that is likely to raise its body temperature above the surrounding air temperature. (1 mark)

(b) *lying on a hot rock*

Here is an example of a behavioural adaptation by an ectotherm to raise its body temperature, above the environmental value. This is why using the terms 'warm-' and 'cold-blooded' are not really satisfactory descriptions. 1 mark awarded.

(c) What evidence from the graph suggests that the bobcat has mechanisms to prevent overheating? (1 mark)

(c) *no matter how the environmental temperature changes above or below its body temperature, the body temperature remains constant at around 40°C*

Reading from the graph as instructed and not pulling the answer from elsewhere, this candidate has spotted that the bobcat can maintain its body temperature constant at 40°C, even if the environmental temperature deviates from this by a substantial amount. A very full answer here which gets 1 mark.

(d) Explain why the bobcat's metabolic rate is greater at 10°C than at 30°C. (2 marks)

(d) *when it gets increasingly cold outside, the bobcat's body loses more heat which must be replaced. This is done by increasing the rate of metabolism to generate this heat.*

When it gets very cold, we know that we bring into play corrective mechanisms to compensate. One example is that we shiver. However, you are actually given the corrective mechanism which the bobcat uses, increased metabolism, and asked to explain why this happens. This candidate has given another excellent response for both marks to be awarded.

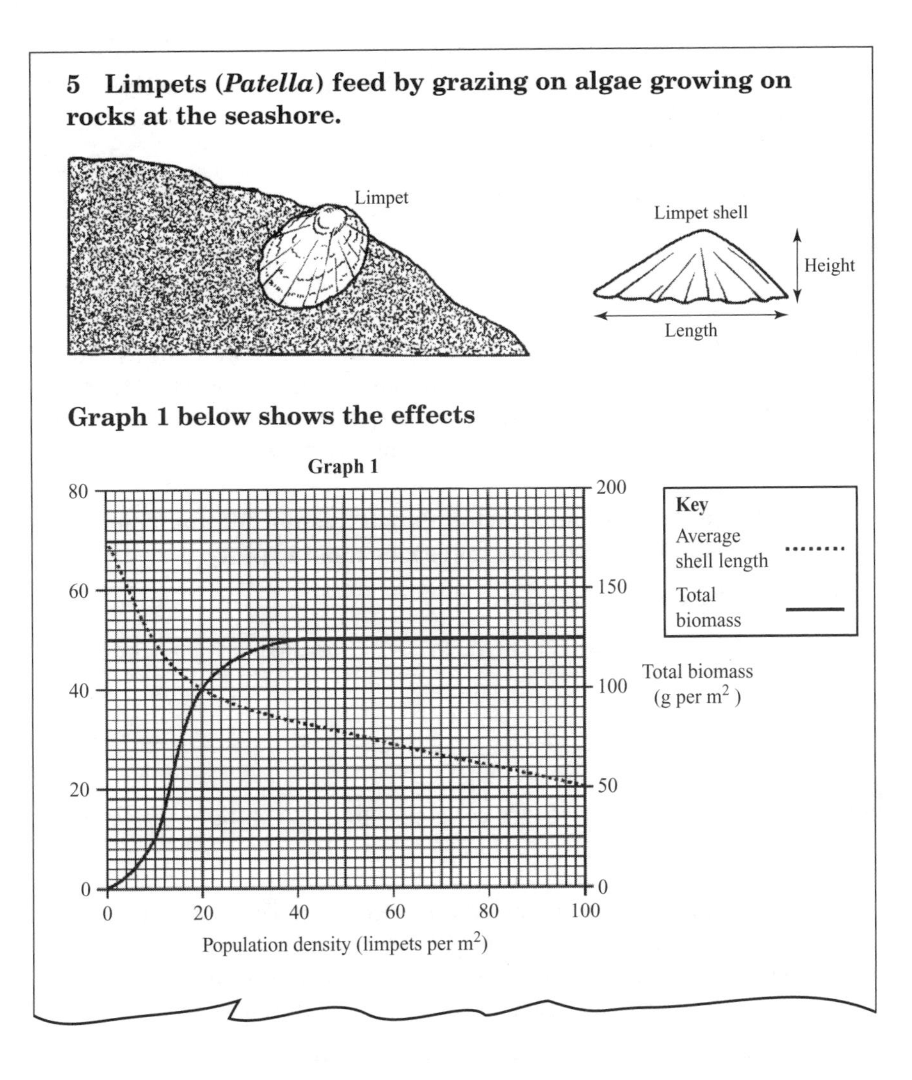

Here is an example of a question which is mainly about data-handling. It touches very slightly on competition but its aim is to test your fluency in reading information and then manipulating it in various ways. Often these questions can spread over several pages because of the graphs and/or tables but do not be daunted by this. Take the question stage by stage. Even if you can't answer one part, that does not mean you can't achieve good marks in the other sections. Usually they are constructed in such a way that answers to one section are not necessarily dependent on getting a previous section correct.

Let's go through this question stage by stage.

(a) What is the total biomass at a population density of 10 limpets per m²?

g per m² (1 mark)

(a) 25

This graph is slightly more complex than some others in that it has two sets of data plotted on the same grid. Each set of data has its own y-axis, left and right of the grid. You must be careful to read the appropriate values linked to the key given. This answer is correct and would obtain the mark.

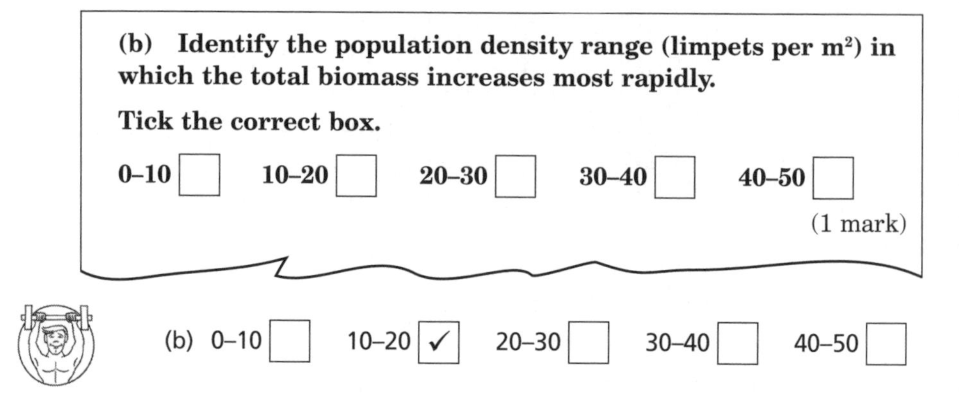

(b) Identify the population density range (limpets per m²) in which the total biomass increases most rapidly.

Tick the correct box.

0–10 ☐ **10–20** ☐ **20–30** ☐ **30–40** ☐ **40–50** ☐

(1 mark)

(b) 0–10 ☐ 10–20 ☑ 20–30 ☐ 30–40 ☐ 40–50 ☐

To do this question, you need to consider the increase in biomass for each of these ranges. For example, 0–10 limpets per m² would give a rise 0 to 25 g per m², then 10–20 limpets per m² would give a rise 25–100 g per m². When you do this for each range you will see that this student has selected the correct answer. 1 mark awarded.

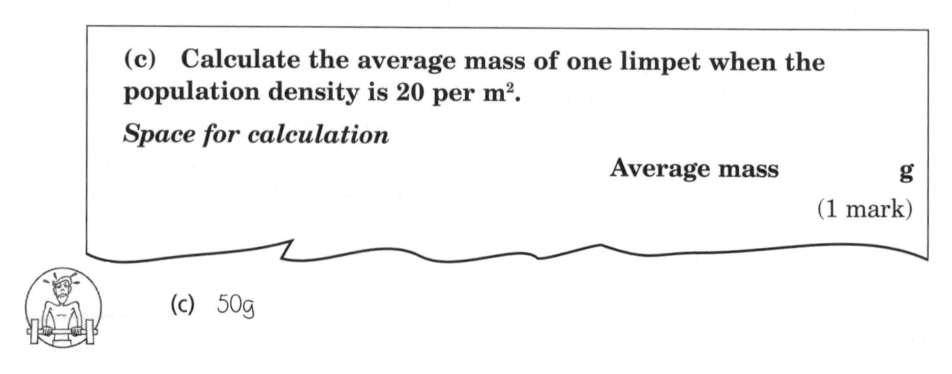

(c) Calculate the average mass of one limpet when the population density is 20 per m².

Space for calculation

Average mass **g**

(1 mark)

(c) 50g

Calculating averages should be straightforward, requiring you divide a total by the number of values. Averages are often used to give an indication of the overall value for some variable. For example, the average heartbeat for an adult male might be around 70 beats per minute. In this question, you need to read the graph carefully but notice how it has been constructed to make this easy. By looking up from the x-axis at the value 20 limpets per m^2 you get a value on the y-axis for total biomass of 100 g per m^2. Dividing 100/20 give you a value of 5 g but this student somehow has been slightly careless perhaps in writing 50 g. You need to watch yourself in this type of situation, making sure all the zeros are in and out appropriately. No marks awarded.

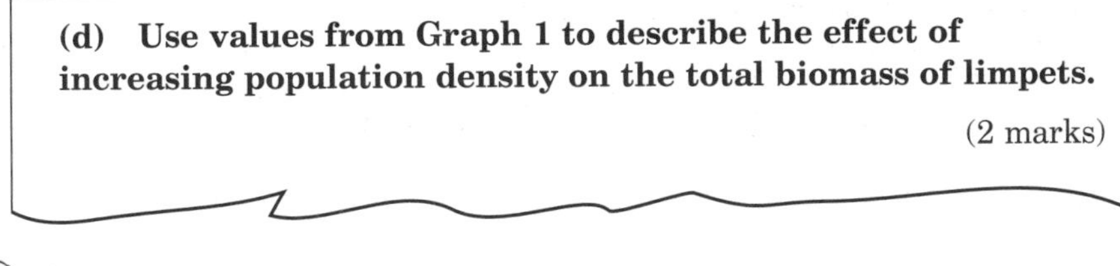

(d) Use values from Graph 1 to describe the effect of increasing population density on the total biomass of limpets.

(2 marks)

(d) *increasing population density increases the total biomass of the limpets*

Questions such as this test you ability to use the data given. It is almost never enough to simply state something increases in these data-handling questions because you need to use the figures given. This student has made a correct answer but only in part and is far too short given the weighting and spacing. Compare it with this answer: *as the population density increases from 0 to 40 limpets per m^2, the total biomass increases from 0 to 125 g per m^2 after which the total biomass remains constant at 125 g per m^2.* No mark awarded.

(e) Explain how intraspecific competition causes the trend in average shell length shown in Graph 1. (1 mark)

(e) *competition between the same species*

Intraspecific competition is indeed competition between the same species but this answer does not explain the trend in average shell length as asked. As always, remember that explanations require more text than a few words to get the mark. A better answer might be *as the population density increases, there will be more competition among the limpets for the same resource so shell length decreases.* No marks awarded.

(f) The table below shows information about limpets on shore A which is sheltered and on shore B which is exposed to strong wave action.

Graph 2 below shows the effect of wave action on limpet shell index.

Limpet shell index = shell height/shell length

Shore A		Shore B	
Shell height (mm)	Shell length (mm)	Shell height (mm)	Shell length (mm)
16	52	9	21
19	54	11	26
20	55	14	31
21	56	16	34
22	57	17	35
23	58	17	36
26	60	–	–
Average = 21	Average =	Average = 14	Average =

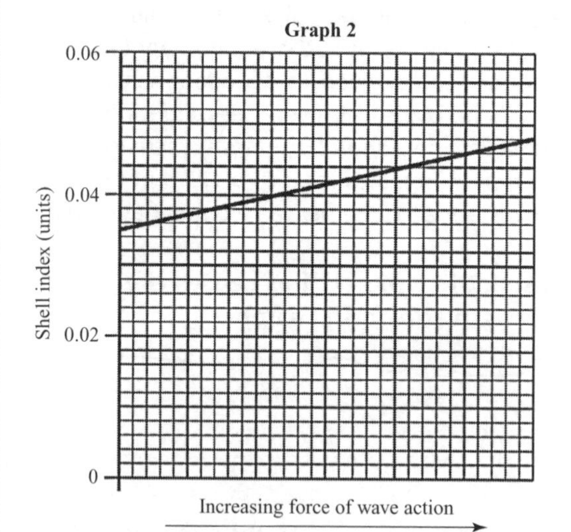

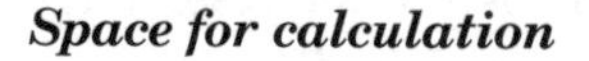

(i) ***Complete the table*** **by calculating the average shell length of limpets on both shores.**

Space for calculation

(1 mark)

(i)

Shore A		Shore B	
Shell height (mm)	**Shell length (mm)**	**Shell height (mm)**	**Shell length (mm)**
16	52	9	21
19	54	11	26
20	55	14	31
21	56	16	34
22	57	17	35
23	58	17	36
26	60	–	–
Average = 21	Average = 56	Average = 14	Average = 31

A simple calculation, totalling each column and dividing by the number of values. Notice the average for Shore B's shell length is actually 30·5 mm but it is allowed to be rounded up as this student did. 1 mark awarded.

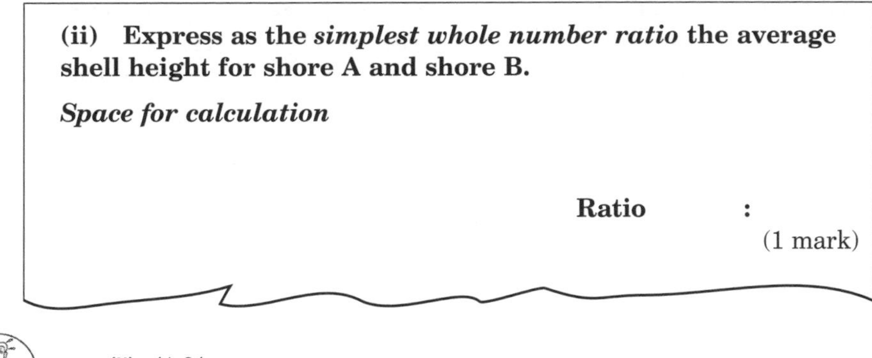

(ii) Express as the *simplest whole number ratio* the average shell height for shore A and shore B.

Space for calculation

Ratio :

(1 mark)

(ii) 14:21

Ratios surface in different parts of the course, for example, in genetics. You need to be able to handle these correctly. Notice this question makes it clear that you must express this ratio as the 'simplest whole number' which you should do anyway as a matter of good practice. This student has made two very common errors. First, the answer is not a simple whole number ratio since each value can be reduced further to 2 and 3 by dividing each by 7. Also, the student has reversed the order of the shores so that ratio effectively is for the ratio of shore B to shore A, not the other way round as asked. The correct answer is 3:2. No marks awarded.

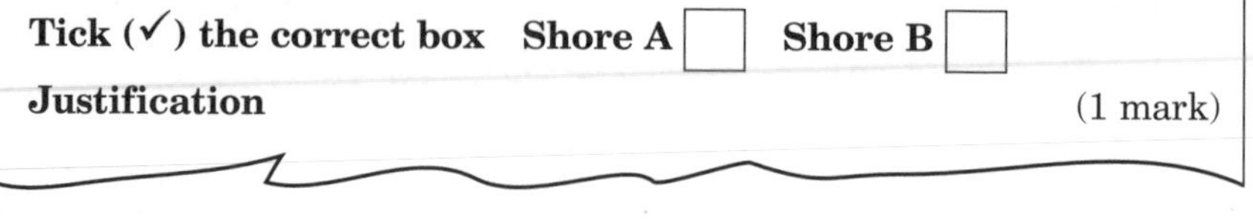
(iii) A limpet shell collected on one of the shores had a length of 43 mm and a height of 20 mm. Use Graph 2 to identify which shore it came from and justify your choice.

Tick (✓) the correct box Shore A ☐ Shore B ☐

Justification (1 mark)

(iii) Tick (✓) the correct box Shore A ☐ Shore B ☑

the shell index is 0·465

This section requires both answers to get the mark. The student here has given the correct shore but not a full enough justification. Simply stating the shell index is not sufficient. The wording does say 'Use Graph 2' therefore you need to use this in some way. You can see from Graph 2 that the calculated shell index is plotted well along the x-axis (increasing force of wave action) so must come from an area on the seashore where wave action is high. A justification might read *the shell index of 0·465 indicates an area of the seashore with a high level of wave action.* No mark awarded.

3 Extended Responses

General advice

Structured extended response questions

Unstructured extended response questions

GENERAL ADVICE

Each of the questions in Section C of the National Examination is worth 10 marks. They are of two types: the first is structured, where the question is broken down into a number of subheadings each with an appropriate allocation of marks. Here is an example:

> **Discuss the factors leading to variation under the following headings:**
>
> **(i) independent assortment** (4 marks)
>
> **(ii) crossing over** (3 marks)
>
> **(iii) mutation** (3 marks)

The second type of extended response question is unstructured, where all the marks are linked to a single question with no subheadings. Here is an example:

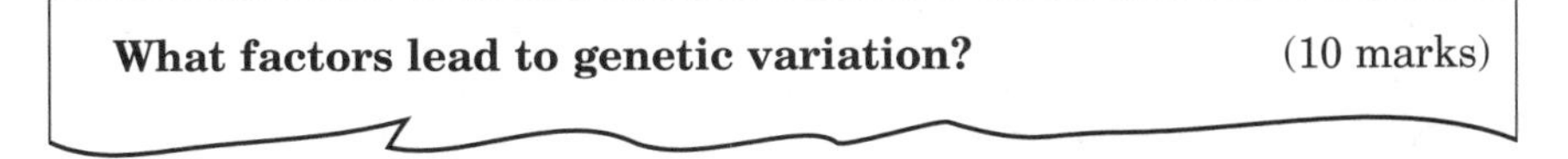

> **What factors lead to genetic variation?** (10 marks)

It is common knowledge to experienced teachers that this section is often found to be the most challenging for candidates, not through lack of knowledge but often by not collating relevant facts and presenting them coherently. This can come only with practice and feedback so you would do well to spend time working on your extended responses. Here are a number of guidelines to help you before we look at some exemplars.

- Sometimes written answers can be improved by relevant diagrams, bullet points, flow-charts, formulae or equations. If you write something which is potentially ambiguous, a diagram, formula or equation may make clear that you really understand the point. If you do decide to use a diagram, make sure it is large and well labelled with non-arrowed lines ending exactly on the structures being identified. The diagram must contain all the information required by the marking instructions. Similarly, bullet points have to have the information required by the marking instructions. You should be aware that you might not get the mark for coherence here if the information is not in paragraphs and only in the form of a diagram, bullet point list or flow-chart.
- Before starting your answer, particularly in response to the unstructured question, make a short plan. Candidates often feel they cannot spare this time, yet experience shows that a few minutes spent on this will result in a better constructed answer. Such a plan reduces the risk of omitting relevant facts and ideas or forgetting to put them in the opening paragraphs and having to add them in paragraphs where they do not belong.
- Be careful to pace yourself in this section of the examination. There is a law of diminishing returns for the time you spend on a single question – so spread your effort over both. If you were to devote all your time to just one question, even with a perfect answer, you could only score 10 marks out of 20. In other words, it is easier to pick up more marks by doing both questions. It is important to be able to have an awareness of time on this section. As a rough guide, spend around 15 minutes on each question. You might find that you have more time if you have answered the previous sections, particularly the multiple-choice, quickly.
- In the unstructured questions, 2 marks are allocated, 1 for coherence and 1 for relevance. We will discuss these separately.
- Coherence really means the essay hangs together well and flows logically from one point, or paragraph, to the next. An obvious way to achieve this is to divide your answer into paragraphs which deal with the question in logical chunks. For example, a question such as '*Discuss the role of enzymes as catalysts*' could be answered under subheadings such as '*Catalysts*', '*Properties*', '*Specificity*' and so forth. Dividing up your answer in this way helps ensure the flow of ideas and makes it more likely that you are awarded this mark. Sometimes, you can take the subheadings from the actual question, as here for example: '*Give an account of the role of the*

pituitary gland in controlling normal growth and development and describe the effects of named drugs on fetal development'. This could be answered under the headings '*Role of the pituitary*' and '*Effects of named drugs on fetal development*'. It is very important that related material should be kept together. In general, you need to aim to have at least five correct points to be able to get the mark for coherence. The examples which follow will make this clear.

- Relevance is a major problem for many candidates. The inclusion of irrelevant information, even if it is itself correct, will gain you no marks. If, for example, you are asked a question about respiration in animal cells and you give information about photosynthesis, even if correct, this will be deemed irrelevant. In other words, stick to the question and try not to deviate from it. You need to aim for at least five correct and relevant points to be able to get this mark. The examples which follow will make this clear.
- When you start the paper, don't waste time going to this section to see what the questions are. This is not good practice. It is much better to work in a methodical way through the sections, building up your confidence and freeing up ideas in your mind before attempting to write the extended responses.
- It is very important here, as has been mentioned several times, not to compartmentalise your knowledge. If, for example, you are asked to: '*Describe the role of water in biological systems*' you might be daunted to start with but in fact you know a great deal of highly relevant information. Think of solvent, temperature control, photosynthesis, lubrication, transport and so forth.
- Take a few minutes here to select questions which are most suited to what you feel you know best. This may seem obvious but examiners will tell you how often candidates start one question then decide, after wasting time on this, to start the other. A little reflection before you put pen to paper is time well spent.
- A good scientific answer will be a combination of knowledge and expression, and this requires practice. You can hope to improve only by repeated attempts, thus building up your confidence, timing and expertise.

STRUCTURED EXTENDED RESPONSE QUESTIONS

Here are two typical examples of structured extended response questions.

1 Discuss meiosis under the following headings:

(i)	**importance of meiosis;**	(2 marks)
(ii)	**independent assortment of chromosomes;**	(4 marks)
(iii)	**formation of chiasmata.**	(4 marks)

2 Write an account of photosynthesis under the following headings:

(i)	**role of leaf pigments;**	(6 marks)
(ii)	**ecological importance.**	(4 marks)

Sometimes these questions are broken up into two or three subsections with the marks often unevenly spread. It is very important that you reflect this in your answers by writing more for heavier weighted subsections and less for subsections which carry fewer marks. Also, it is not necessarily the quantity of words but the quality and relevance of the points which is important. Although there is no allocation of marks in these questions for coherence and relevance, these important aspects of good extended writing still apply. Use the subheadings given to help remind you what each subsection is about. When you move onto a new page in your examination answer book, you may become mentally detached from the actual question and can easily wander off what is being asked or repeat something you wrote earlier. By using this simple device of identifying each subsection, this error is less likely to happen.

Let's go through some examples from Section C of actual National Examinations and see what makes for good and bad structured extended responses.

1 Write notes on each of the following:

(i)	**the structure of the plasma membrane**	(3 marks)
(ii)	**the structure and function of the cell wall**	(3 marks)
(iii)	**phagocytosis**	(4 marks)

Here is a good pupil's answer.

(i) *The diagram below shows the double layer of phospholipids and the random distribution of protein molecules throughout which form a mosaic. Some proteins can form channels as shown.*

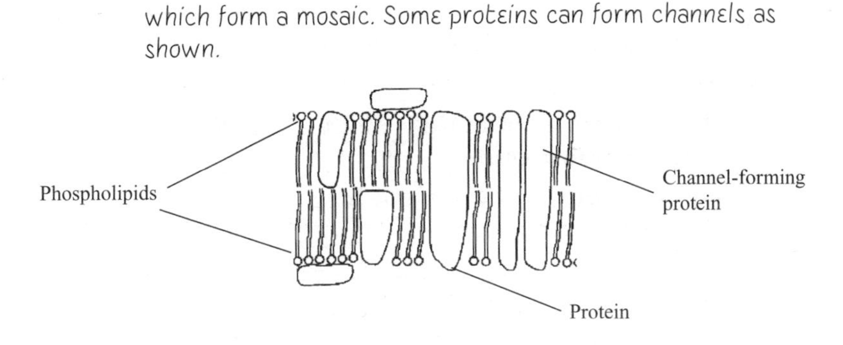

Notice how this pupil has used a very neat diagram to enhance the text. Each structure is correctly labelled with non-arrowed lines and ending exactly on what is being identified. 3 marks awarded here.

(ii) *The cell wall is chemically composed of cellulose which gives the plant cell rigidity and shape. It is fully permeable to water but is strong enough to prevent the cell bursting by osmosis.*

This question, unlike the previous one, requires more than just structure. Function is almost always closely linked to structure. Here the pupil has correctly linked the strong chemical basis of the cell wall to function. 3 marks awarded here.

(iii) *The diagram below shows a phagocyte about to surround some bacteria. These bacteria are then enclosed in tiny membrane-bound sacs. Next, lysosomes become attached to these sacs and empty their powerful digestive enzymes onto the bacteria destroying them.*

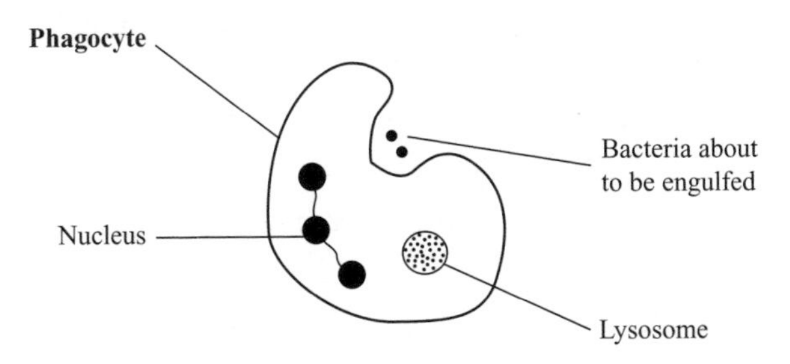

Again see how this pupil has made the answer very clear by using both text and a well-labelled diagram. The sequence of events is set out in the text above the diagram. 4 marks awarded here.

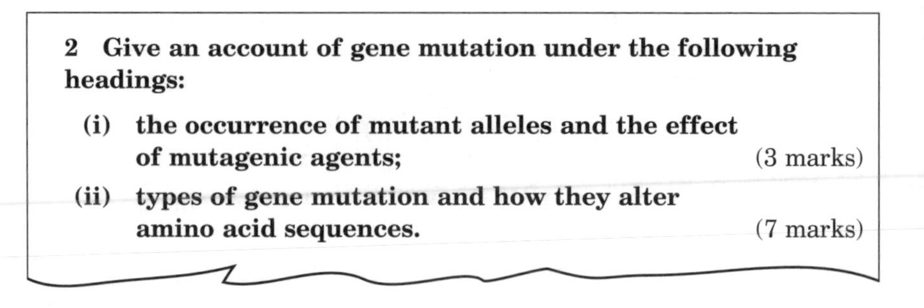

2 Give an account of gene mutation under the following headings:

(i) the occurrence of mutant alleles and the effect of mutagenic agents; (3 marks)

(ii) types of gene mutation and how they alter amino acid sequences. (7 marks)

Here is one poor attempt at this question and we'll see why.

(i) *Down's syndrome is caused by a mutation which means the person can be mentally retarded. An extra chromosome is found in their cells.*

First, the student has not included the subheadings though this of course would not be penalised. Clearly, if you were at all rushed for time, you could simply use the numbers of the subheadings but this will not prompt you to be relevant. Next, the answer is about chromosome mutation, not gene mutation, so is entirely irrelevant to this question, even though it is factually correct. Finally, had the question been on chromosome mutation, this answer would have been insufficiently detailed for the award of 3 marks. No marks awarded here.

Here is a good answer.

(i) *Mutations are random events which happen very rarely. They can be caused by certain chemicals such as mustard gas.*

Notice how this student has used the subheading to help keep the answer on track. Each point is entirely relevant so no time has been wasted writing irrelevant information. If we were to look at how an examiner would mark this answer, each point could be underlined (or ticked) rather like this:

Mutations are random events which happen very rarely. They can be caused by certain chemicals such as mustard gas.

Notice the link between '*cause*' and the example of '*chemicals such as mustard gas*'. 3 marks awarded here.

This is the poor student's attempt at the second part of the question.

(ii) *Gene mutations don't happen very often and can be very harmful. There are four types of gene mutation, substitution, deletion, inversion and insertion. Substitution means one base gets substituted for another, deletion happens when a base gets lost, inversion causes bases to be switched around and insertion happens when a base gets added. Because genes make proteins, mutations can cause the wrong proteins because the amino acids are all wrong.*

This is not a great answer and is unlikely to score more than 3 marks. Having omitted the subheading in part (i), the student has done the same here. The opening sentence '*Gene mutations don't happen very often and can be very harmful*' belongs to the first part of the question so is not relevant here. The four types of gene mutation are correctly stated but notice how the student uses similar words in the same sentences: '*Substitution ... substituted*' *and* '*... genes make protein ... wrong proteins ... all wrong*.' This is not only bad English but also not good Biology. You need to avoid this error, typical of weaker students, by reading over what you have written. It only takes a very short time but it will help you improve the quality of your writing. Genes don't make proteins directly but they do 'code' for them. Any change in the gene for a protein would result in some change(s) in the amino acid chain. The student has not developed this sufficiently. Notice, finally, the number of lines written does not reflect the weighting of this part of the question which is more than double the first part.

Here is an answer from the same good student who got 3 marks earlier.

(ii) *A gene mutation changes the order of the bases in the DNA. This can happen in four different ways.*

1 *Substitution – a base is replaced by another one.*

2 *Deletion – a base gets removed.*

3 *Inversion – two bases get switched round.*

4 *Insertion – a base get added.*

Insertion and deletion mutations cause the whole sequence of bases to be changed after mutation and therefore all the amino acids from this point on are changed in the final protein.

Inversion and substitution only change one triplet and therefore only one amino acid is changed in the final protein.

Notice how this student has spaced out the answer, and cleverly used a bullet point type approach for each of the four types of mutations with good explanations of each. The use of these basic presentation skills enhances a response and makes it easy for the examiner to pick up the points for award. All of this answer is completely relevant to the question with no repetition. 7 marks awarded.

3 Give an account of populations under the following headings:

(i) the importance of monitoring wild populations; (5 marks)

(ii) the influence of density-dependent factors on population changes (5 marks)

An average student's attempt.

(i) *There are several reasons for monitoring populations. Fish for example may be used for food so it is important not to over-fish. Some plants and animals may be in danger of extinction. Pollution in water can be checked by looking at indicators. Some insects, such as aphids, can be pests if their numbers get too large and need to be controlled.*

In this answer, the student demonstrates some knowledge but has not given enough nor linked reasons and examples very clearly. The opening sentence '*There are several ....*' is not necessary and would not get any marks. Look at the first example of fish. It is true that monitoring food stocks such as fish is vital so that a decrease in numbers can be detected quickly and action taken to prevent over-exploitation. This type of presentation is needed for each example but the student has not carried this through. Look at the pollution which has no example of an indicator nor a good link to why indicators can inform about the state of a water supply for example. This answer would probably be awarded 3 marks.

Here is what is needed to obtain full marks.

(i) *The importance of monitoring wild populations*

- *Animals, such a deer, which are used for food cannot be over exploited or their numbers will not be maintained.*
- *Rare animals, such as pandas, are in danger of extinction so their numbers must be carefully monitored.*
- *Mayfly numbers in water are a good indicator of the level of oxygen in water since they won't tolerate pollution which lowers it.*
- *Lichen on trees are very sensitive to air pollution so their numbers will decrease for example if sulphur dioxide levels rise.*
- *The effect of an environmental disaster, such as an oil spillage, can be assessed by looking at changing numbers of sea-life such as birds.*

See how this student has made sure of linking the 5 marks allocated to five distinct points using a bullet point presentation. Also, each point gives a clear example of an organism whose numbers might need to be monitored and why. An examiner would find this a very easy answer to mark and give a full award.

Here is the average student's attempt at the second part of this question.

(ii) *As the numbers of a population increase, the effect of density-dependent factors increases. For example as the numbers of rabbits increase, there will be an increase in the number of foxes. This helps to keep the numbers of rabbits at a constant level. Other factors might be intraspecific competition, which is competition among the same species for something in the environment such as food. More disease happens when the population increases.*

This answer would be unlikely to score more than 3 marks. The student does seem to know what is meant by the question because the opening sentence explains nicely the link between changing population numbers and the changing effect of a density-dependent factor. Notice, however, how this student starts to wander from the question by explaining what is meant by intraspecific competition which is not part of what is being asked here.

To obtain full marks an answer like this is required.

(ii) *The influence of density-dependent factors on population changes*

As the name suggests, the effect of a density-dependent factor is linked to the numbers of a population. As the numbers of the populations increase, so does the effect of the density-dependent factor. For example, as the density of a prey increases, so does the effect of predation. As density of a population increases, so do the parasites which infect that population. Competition for the same limited resource will increase as the population numbers increase. The overall effect of this is to help keep the population numbers stable.

This is a well-constructed paragraph, flowing nicely from the initial statement, explaining the link between density-dependent factors and population changes then moving on logically to examples. It ends with the net effect of the link which rounds off the paragraph very well.

4 Give an account of transpiration under the following headings:

(i) the effect of environmental factors on transpiration rate; (5 marks)

(ii) adaptations of xerophyte plants that reduce the transpiration rate. (5 marks)

Here both sections are equally weighted and you should try to reflect this in your answers.

A weak candidate's attempt.

(i) *Factors in the environment could be light and temperature. If these increase there is an effect on transpiration. Also, if the wind is blowing, it might change transpiration as well.*

This student clearly is lacking in knowledge for this topic. While fragments of the answer appear, the student would be very unlikely to obtain any marks for this response because there is no link between the variables. You must always be able to predict or state what the effect of changing one variable will be on the other. 'As x increases, y decreases' for example. Also, given the weighting, the content here is lacking too much. Remember that 'light' is normally not accepted as a variable, you should always write 'light intensity'.

A good candidate's attempt.

(i) *The effect of environmental factors on transpiration rate*

Factor	Effect of changing factor
light intensity	increase would increase transpiration
wind speed	increase would increase transpiration
temperature	decrease would decrease transpiration
humidity	increase would decrease transpiration
air pressure	increase would decrease transpiration

See how this candidate has used a neat way of presenting the information for this question, using a table with correct column headings so that each factor is correctly linked to the effect of changing it. Tables like this can save a lot of time but still present the examiner with clear evidence of knowledge which is correct and relevant. Also, the candidate demonstrates a good understanding that factors can both increase and decrease and how these different changes can affect the rate of transpiration. Full marks awarded here.

A weak candidate's attempt

(ii) *Xerophytes are adapted to growing in water and have stems and leaves exposed to the air. The body parts have lots of air spaces so they can float and exchange gases easily. They are very delicate plants.*

Elsewhere, the subject of biological terminology has been discussed. Difficult words like 'xerophyte' need to be memorised very thoroughly. This weak candidate has, unfortunately, thought the question was about 'hydrophytes' so no marks are awarded here. It is very hard to pick this up once the initial error has been made, so the candidate continues to write about hydrophytes instead of what was asked.

To obtain full marks, look at this response from a very able candidate.

(ii) *Adaptations of xerophyte plants that reduce transpiration rate*

Xerophytes live in harsh dry conditions such as desert and so have many adaptations to prevent the loss of water.

- *Guard cells are sunk deep into pits so that water vapour gathers and helps stop further evaporation.*
- *Leaves may have hairs and thick cuticles which helps trap water vapour and stop water loss.*
- *Some, such as marram grass, have leaves which can roll to cause build-up of humidity and prevent excess water loss*
- *Some xerophytes, such as cacti, store water in special succulent tissue.*
- *Plants may show a reversed pattern of stomatal opening and closing opposite to that of normal land plants which means stomata are open at night and closed during day.*

This is a first-class answer from a very well-prepared candidate who has learned this theory in-depth. Notice how an explanation of xerophyte is given and linked nicely to the need for adaptations to prevent water loss. The candidate lists, using bulleting, a number of structural adaptations and how these reduce transpiration but notice how a 'physiological' adaptation is also given, that of reverse stomatal rhythm.

UNSTRUCTURED EXTENDED RESPONSE QUESTIONS

Here are two typical examples of unstructured extended response questions.

(i) Discuss the role of meristems in plants. (10 marks)

(ii) How is human growth affected by the pituitary and thyroid glands? (10 marks)

Students usually find these unstructured questions more difficult than structured ones for the obvious reason that there is no break up of the marks into smaller sections which would make it easier to write something for each. A candidate, for example, who had not studied up plant meristems, would be almost totally at a loss to handle question 1 here. Whenever the question has different parts implied by the wording, use these to form subheadings in your answer. This will help you to write in suitable paragraphs and obtain the mark for coherency quite easily. These questions definitely lend themselves to forming a brief plan and you will find that this helps enormously. Sometimes, just the mere act of writing down key words actually triggers the memory of other related ideas as well. Try this for yourself – it really does work. A brief plan also helps the flow and logic of your presentation. Look at this plan for essay 1 above:

brief definition – look like – form new cells – mitosis – all other plant tissues – active growing points – stem and root meristem – lateral meristem – differentiation – xylem and phloem – new leaves, roots, shoots – primary growth – where found

Next you might add numbers to indicate paragraphs so that the terms all link together. For example it might start off, as shown below, for the first paragraph.

1 brief definition – 1 look like – 1 form new cells – 1 mitosis – 1 all other plant tissues – active growing points – stem and root meristem – lateral meristem – differentiation – xylem and phloem – new leaves, roots, shoots – primary growth – 1 where found

See how the candidate remembered something later '*where found*' and so immediately wrote this down at the end but numbered it 1 for inclusion in the first paragraph. This is a very good habit. Don't make the fatal mistake of thinking 'I'll remember it' – you won't! Write any new related ideas down as soon as they come into your mind before they disappear as your concentration moves onto some else. It doesn't matter if it is out of context, you can sort that later. At the end, score out your plan so the examiner knows not to read this.

Let's go through some examples from Section C of actual National Examinations now and see what makes for good and bad unstructured extended responses.

1 Give an account of chloroplast structure in relation to the location of the stages of photosynthesis and describe the separation of photosynthetic pigments by chromatography. (10 marks)

Here is an average pupil's attempt at this question.

Chloroplasts are found in green plant cells and carry out photosynthesis. They have layers of membranes called grana inside where chlorophyll is found. This is where photosynthesis takes place in the light-dependent stage.

A suitable plant for extracting pigments is nettle. The leaves are ground up with sand and then filtered to give an extract which is placed on a piece of special paper many times to make a spot. A solvent is used and this runs up the paper separating the spot into the different pigments, carotene, xanthophyll and chlorophyll.

This question is really in two distinct parts, one is on the structure and function of the chloroplast and the other on an experimental procedure to separate the pigments in a leaf. Notice, this student did not give subheadings though paragraphs were used which is acceptable for coherence. The information for each part must be properly grouped together and there must be at least five correct and relevant points spread between the two parts with at least two correct points from each and the fifth from either. Not all of the four pigments you need to know at Higher Grade level have been given. An examiner reading this average pupil's reply might underline the correct points as follows.

Chloroplasts are found in green plant cells and carry out photosynthesis. They have <u>layers of membranes called grana</u> inside where <u>chlorophyll is found</u>. This is where photosynthesis takes place in the <u>light-dependent stage</u>.

A suitable plant for extracting pigments is nettle. The leaves are ground up with sand and then <u>filtered to give an extract</u> which is placed on a piece of special paper <u>many times to make a spot</u>. A solvent is used and this runs up the paper separating the spot into the different pigments, carotene, xanthophyll and chlorophyll.

For the first paragraph, three relevant points are made and grouped together. In the second paragraph, two relevant points are made and grouped together so the coherence mark would be awarded. Notice the important thing here is the grouping of the relevant points, in other words, cohering them.

For the relevance mark, the answer must not contain references to other structures such as the mitochondria. It must contain also contain five correct and relevant points, as outlined above. The relevance mark would be given here as well.

No plan was made by this student which is usually not a good idea. If you have time, make a brief outline of key ideas and terms.

In total this answer would receive around 7 marks.

Here is a suggested answer which would score very high marks.

Plan

double-membrane/carbon dioxide/water/stroma (ATP etc)/
starch/lamellae/grana light stage/chlorophyll/dark stage

flow diagram/grinding leaves/filtration (why)/spotting (repeat)/
solvent run/4 extracts

Structure of chloroplast

The following diagram shows the basic structure of the chloroplast as well as where the different stages of photosynthesis take place.

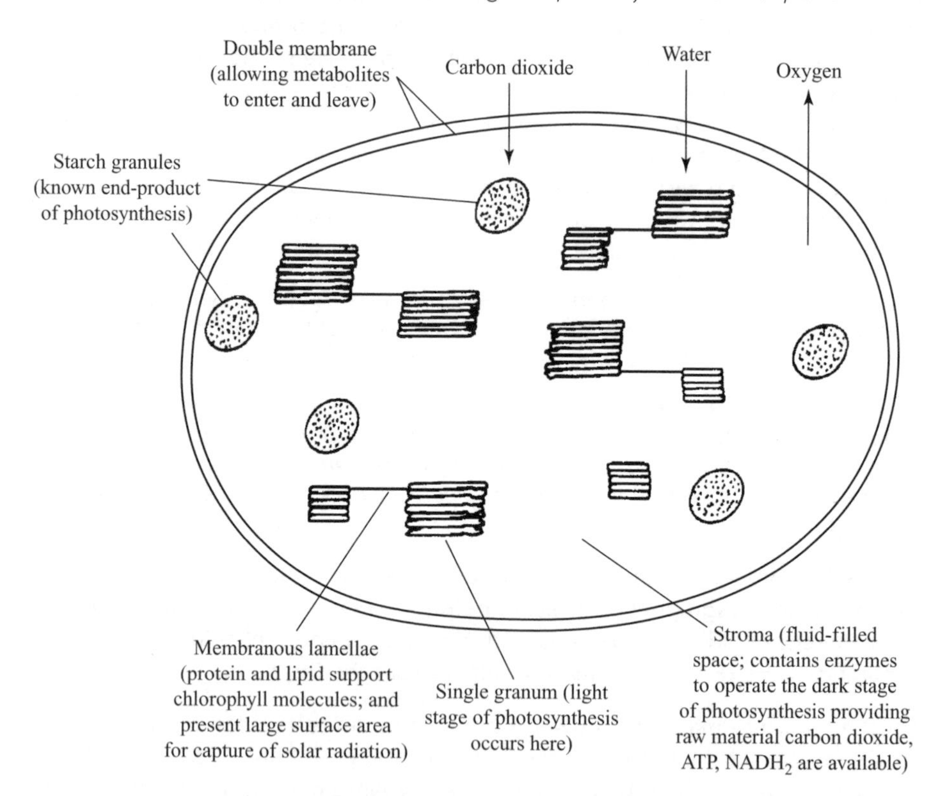

Notice that this student has made an overall plan but divided it up to correspond with each of the two sections. By doing this, it is less likely that you will mix up information or miss out facts. As you move through the plan or even once you have started writing, if you remember something you want to add, stop and write in into the plan immediately. At the end, you can tick off the points to make sure you have missed nothing out.

See how this student has made use of the technique of drawing instead of a lot of writing to convey information. After a well-constructed, short plan, the rest is easy. All the important structures are shown and well-labelled with non-arrowed lines. In addition, the student has indicated what happens in each part of the chloroplast which is related to photosynthesis. Notice that the diagram does not have to be artistic so long as it is anatomically accurate, in other words, it shows the internal structure clearly and unambiguously. The fact that the student may have gone beyond what was asked indicates the depth of knowledge possessed. However, the student has not written up the answer in paragraph form and may lose the mark for coherence here.

Separation of photosynthetic pigments by chromatography

Separating pigments by chromatography requires several stages as follows:

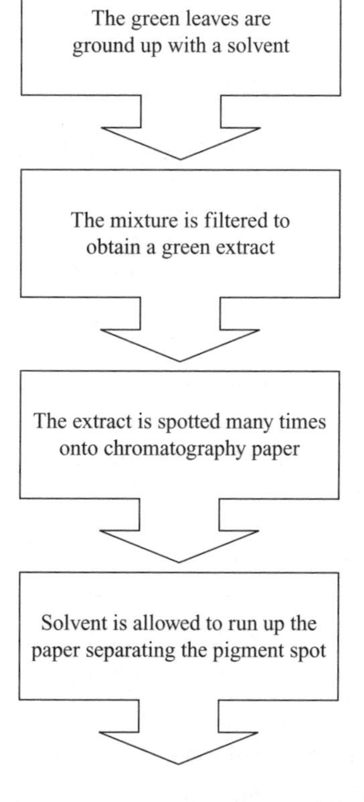

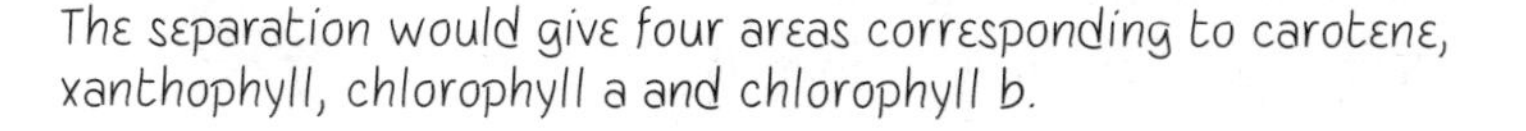

This student has answered the question by using a flow-chart to show the sequence of actions taken to produce the chromatogram. The final sentence shows clearly the student knows exactly what extracts would be produced from this procedure. Notice also the use of subheadings taken from the wording of the question itself and that more than enough appropriate and correct information is given to obtain the mark for relevance but, as stated earlier, the mark for coherence may not be awarded.

Think about using diagrams in this way, particularly when the question relates to structures, which are often easier to explain with a well-labelled drawing but try to include paragraphed text as well to ensure you get the mark for coherence.

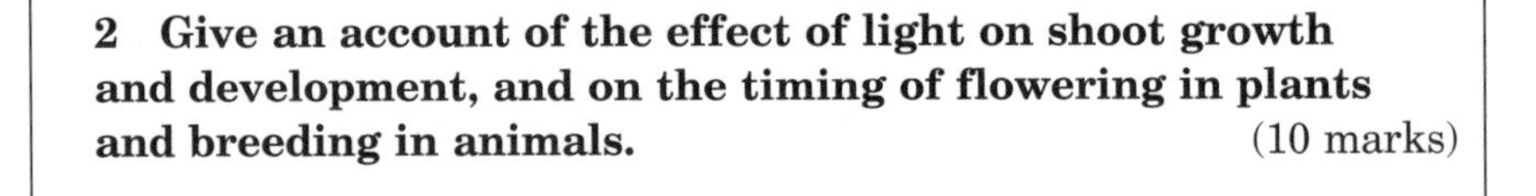

2 Give an account of the effect of light on shoot growth and development, and on the timing of flowering in plants and breeding in animals. (10 marks)

Here is a very weak attempt at this question.

Plants need a certain amount of light to flower. When the day is too long some plants won't flower. Light also makes shoots bend, which is called photoperiodism. If plants don't get enough light they become weakened. Animals also need a certain amount of light for breeding. Some animals breed only in autumn and others only in springtime. If the length of day is too short, it may cause some plants not to flower.

This student shows very clearly a complete lack of knowledge of this topic with a bad error in using 'photoperiodism' instead of 'phototropism'. Also, it is obvious that there has been very little practice in the style of answering unstructured extended responses. Information is mixed up with no flow or logic. No paragraphs are formed nor are any subheadings used. Even a rudimentary plan is missing. In terms of quantity and quality of information, relevance and coherence, this answer would almost certainly get no marks at all.

Let's now consider how to tackle this question, which is quite challenging. First, divide it up into three obvious sections from the stem of the question. These are the effect of light on:

1. shoot growth and development
2. the timing of flowering plants
3. the timing of breeding in animals

Having done this, already the answer starts to take shape. The headings help you focus on the theory needed for each section and make it less likely for you to wander.

Next, construct a plan for each section. As already explained, it is a good idea to try and do this at the start for the whole answer.

phototropism/IAA/light and dark sides/cell growth/no light

photoperiodism/explanation/long-day plants/short-day plants

animals and long-day/short-day/why

Now it only remains to flesh out these trigger points into three paragraphs, with the appropriate headings. Let's see how an able candidate did just that.

Shoot growth and development

Plants grow in response to the direction of light, a process called phototropism. This is brought about by different concentrations of the hormone IAA so that more is present on the side which is in darkness compared with the side being illuminated. The increased IAA on the dark side causes the cells to grow faster than those on the light side and so the shoot bends towards the light. If no light is present at all, the shoot will growth vertically upwards but any leaves which are formed will be very yellow.

Timing of flowering in plants

The effect of daylength on flowering in plants is called photoperiodism. Some plants, such as lettuce, are long-day while others, such as chrysanthemum are short-day plants. Long-day plants only flower when the hours of light go above a certain value while short-day plants only flower when the hours of light fall below some critical value.

Timing of breeding in animals

Animals which breed when the hours of daylight go above some critical value are long-day while those which breed when hours of daylight fall below some critical value are short-day. Long days occur in spring/summer so it likely that animals such as birds will breed then while short days occur in autumn/winter when animals such as deer breed. This means the offspring are born when environmental conditions are suitable.

This is a very robust answer to this challenging question. Each section is dealt with separately with no repetition of information or information placed wrongly. Paragraphs are formed with good headings taken from the question. Within each paragraph the facts are well grouped together and flow logically. There are at least five correct points with at least one correct point appearing in each section. There is no irrelevancy by giving information, whether correct or not, which does not bear on the question. The marks for relevancy and coherence would be awarded here along with 3, 3 and 2 respectively for each section. Notice also you don't know exactly the weighting the SQA give for each section. You can judge from experience and practice what the likely mark allocations will be and write accordingly.

3 Give an account of the principle of negative feedback with reference to the maintenance of blood sugar levels. (10 marks)

This essay would cause some students problems because they find feedback difficult. The concept of feedback is very important, not only in biological systems. It is vital to keep changing variables within narrow limits. As an analogy, consider the thermostat which controls an oven. Once you set the temperature at some value and switch on the heating element, the thermostat constantly checks to see if the temperature is rising above the set point, in which case it switches off the heat, or below the set point, in which case it turns the heat back on. Or consider a driver in a restricted zone of say 30 mph. The driver will not be able to drive the car at exactly this speed but will rise above and below it by small amounts and will have to make constant adjustments to slow or speed up the car.

Let's see how this average student tackled the question.

Negative feedback makes sure the body does not lose control of its sugar levels. If this happened, you could become quite ill and possibly become unconcus. You need to keep blood sugar level constant to stop this happening. Once the level rises above a certain value, the body can detect this and make changes to bring it back to normal. Any changes in sugar level are picked up by the pancrease which makes insulin to stop the level getting too high. The pancrease also makes another hormone called glucagen which brings the sugar level back up to normal again when sugar levels fall. This is called homeostasis.

This student knows some information but has not presented it well. Here are some of the issues which make this answer worth possibly only 6 marks.

- There are no paragraphs or subheadings taken from the question which means the coherence mark cannot be awarded.
- No planning has been done which would have helped avoid some of the errors described below.
- Facts are disjointed. For example, the last sentence '*This is called homeostasis*' should really appear right at the start with a clear explanation of what negative feedback actually means.
- There are some spelling errors '*unconcus*' instead of '*unconscious*', '*glucagen*' instead of '*glucagon*' and '*pancrease*' instead of '*pancreas*' but these would not be penalised. The examiner will be sympathetic if you make a wrong spelling providing it is obvious what you mean. However, watch words like glycogen and glucagon which are similar.
- See how this student writes 'you' often. While this does not get marked down, it is not good practice in an extended response to personalise the writing as if you were talking to the examiner. Instead of '... *you could become quite ill* ...' for example, it is much better to rearrange the words to something like '... *a person could become quite ill*'.
- If you get into the habit of reading over your own answers you will avoid the repetition evident in this answer. See how many times the student uses the words '*sugar levels*'. A good tip here is to invite your peers to read your answers. You'll be surprised how useful this technique of 'peer-marking' is. They are often very harsh critics but you'll pick up on issues which may not be obvious to you!
- There is almost but not quite enough correct information in this answer, particularly in the second part dealing with the maintenance of blood sugar levels, to obtain the relevancy mark.

Let's write an answer which will include this average student's answer but improve on it to get full marks.

Plan

definition/diagram/explanation/return to normal
glucose levels in blood/pancreas/rising + insulin/glucose to glycogen/liver/
falling + glucagon/ glycogen to glucose

Principal of negative feedback

Negative feedback is a mechanism for keeping the body's internal conditions constant. This mechanism can be shown by the following diagram:

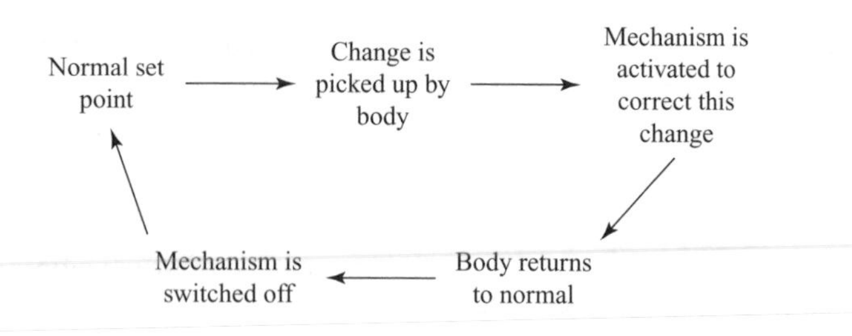

Maintenance of blood sugar levels

The concentration of glucose in the blood must remain constant. An increase above normal is picked up by the cells of the pancreas which release the hormone insulin. This causes the excess glucose to be changed into glycogen to be stored in the liver. When levels fall below normal, the hormone glucagon is released from the pancreas. This causes stored glycogen to be changed back to glucose again. In this way, the levels of glucose in the blood are kept within narrow limits.

This is an excellent response to the question. See how the student uses a plan, paragraphs and subheadings taken from the question to make sure the answer flows, there is no repetition of facts and each paragraph has correct points grouped together. While there are other facts the student could have included, for example the effect of insulin on the permeability of cells, the answer is sufficiently comprehensive to obtain the maximum 8 marks for content as well as both marks for coherence and relevancy.

Notice the use of a diagram in the first section but only after a concise definition of negative feedback has been given. This is a neat summary of the flow between the points, but could have been conveyed equally well by text alone. However the diagram does make the linkage easy to see.

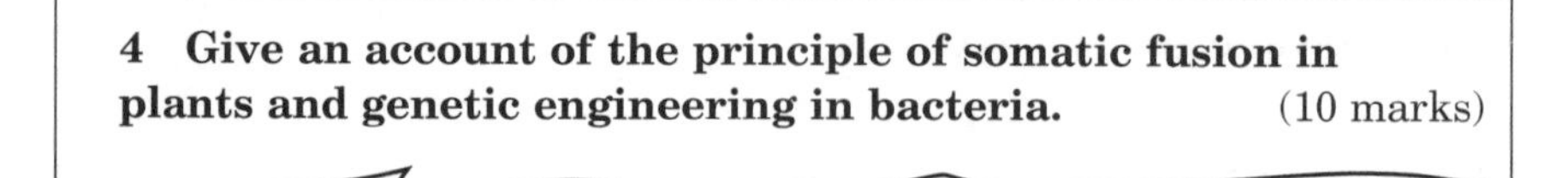

4 Give an account of the principle of somatic fusion in plants and genetic engineering in bacteria. (10 marks)

This essay links two topics in the section on Artificial Selection from the second unit of the course. Somatic fusion is often used to enable two different plant species which, by definition, cannot cross successfully, to combine in the laboratory. The technique allows growers to produce new varieties of plants which are a mixture of desirable characteristics of different species. Genetic engineering enables particular genes from one species to be inserted into an entirely unrelated species which then might, for example, produce a human hormone which it would not do normally.

Here is a good candidate's answer for this question.

Plan

Somatic fusion: *incompatible species combined; plant cells; cell wall digestion; enzyme; cellulase; protoplast; fused; new plant*

Genetic engineering: *locate desired genes; probes; enzymes; endonuclease; ligase; plasmid use; modified bacteria reproduce; rapid; example of product*

Somatic fusion

Somatic fusion allows scientists to force two species which normally could not breed to combine to produce a new variety. For example, using plant cells from a potato and a tomato, it is possible to make a new plant, called a pomato, which has good features of both plants. Such a plant would flower but not produce fertile seeds. In order to fuse the cells, their cell walls have to be digested by an enzyme called cellulase. This produces structures called protoplasts which can be made to combine, maybe using electricity, to make a somatic hybrid. This hybrid can then be made to grow into a completely new type of plant.

Genetic engineering

The first stage in genetic engineering makes use of a genetic probe to find the required genes in the donor organism. The genes are then removed from the donor organism using an enzyme called an endonuclease. Plasmids, tiny rings of genetic material found in bacteria, are cut open using a different endonuclease so that the genes from the donor cell can be inserted. The inserted genes are sealed in place using another type of enzyme called a ligase. The changed plasmid is then put back into the bacterial cells and the modified bacterium allowed to reproduce. Bacteria reproduce every twenty minutes so, very quickly, a culture of modified bacteria will be produced. These bacteria may make human insulin for example.

This is an excellent answer and well planned answer which would obtain full marks.